INTERFACE CULTURE

HOW NEW TECHNOLOGY TRANSFORMS THE WAY WE CREATE AND COMMUNICATE

STEVEN JOHNSON

BASIC
BOOKS

A Member of the Perseus Books Group

Interface Culture was originally published in hardcover by Harper San Francisco in 1997. The Basic Books edition is reprinted by arrangement with Harper San Francisco, a division of HarperCollins Publishers, Inc.

Published by Basic Books,
A Member of the Perseus Books Group

Designed by Laura Lindgren

Library of Congress Cataloging-in-Publication Data
Johnson, Steven.
    Interface culture : how new technology trans-
forms the way we create and communicate /
Steven Johnson.—1st ed.
        p.   cm.
    Includes index.
    ISBN 0-06-251482-2 (cloth); ISBN 0-465-03680-5 (pbk.)
    1. Information technology—Social aspects.   2.
Information society.   3. Communication and cul-
ture.   I. Title.
T58.5.J64   1997
303.48'33—dc21   97-17596

DHSB   02 03 04 05   15 14 13 12 11 10 9 8 7 6 5

For my parents

CONTENTS

ELECTRIC
SPEED

This book is an extended attempt to think about the object-world of technology as though it belonged to the world of culture, or as though those two worlds were united. For the truth is, they have been united all along. Was the original cave painter an artist or an engineer? She was both, of course, like most artists and engineers since. But we have a habit—long cultivated—of imagining them as separate, the two great tributaries rolling steadily to the sea of modernity, and dividing everyone in their path into two camps: those that dwell on the shores of technology and those that dwell on the shores of culture. The opposition colors much of modern thought. (Even the human brain is now seen as having a lobe for artists and a lobe for engineers.) But it is as false as the genetic separation between human and ape. As was true for the triumph of Darwinism in this century, it has taken a combined effort of art and science to make this error visible.

This is not the first book to proceed from these initial conditions. In fact, the argument here belongs—somewhat grudgingly—to a larger movement, one that is only now coming into its own. You need only look around you to see the evidence: the word *technoculture* appears on every other page of *Wired* magazine, and the literature departments that once looked enviously to Paris for the latest bon mot now get their source materials straight from Silicon Valley. John Brockman announces the existence of a "third culture," populated by complexity theorists and multimedia savants. Self-help gurus like Anthony Robbins explicitly describe their guides to personal fulfillment as "technologies," while computer-animation companies like Pixar conjure up entire movies out of binary code. Any professional trendspotter will tell you that the worlds of technology and culture are colliding. But it's not the collision itself that surprises—it's that the collision is considered news. You'd think the life of Leonardo da Vinci or Thomas Edison would be enough to convince us that the creative mind and the technical mind have long cohabited. Alas, the doyens of technoculture are too busy proclaiming the Internet "the greatest thing since the invention of fire" to contemplate the great revolutionaries of the past. The digital world may be jacked in, booted up, and wired for sound, but it has a tin ear for history.

There's a funny thing about the fusion of technology and culture. It has been a part of human experience since that first cave painter, but we've had a hard time seeing it until now. When James Joyce published *Ulysses* in 1922 and revolutionized all of our expectations about how books should work, was he so different from Gutenberg himself? You couldn't see it at the time, but Joyce was a highly skilled technician, tinkering around with a book-machine, making it do things it had never done before. His contemporaries saw him as an artist (or a pornographer, depending on who you talked to), but from our vantage point, he could just as easily be a programmer, writing code for the printing press platform. Joyce wrote software for hardware originally conjured up by Gutenberg. Reverse the angle, and the analogy holds as well: Gutenberg's reworking of the existing manuscript technology of quills and scribes was a creative act as profound as Molly Bloom's final monologue from *Ulysses*. Both innovations came from startling imaginative leaps, and both changed the way we look at the world. Gutenberg built a machine that Joyce souped up with some innovative programming, and Joyce hollered out a variation on a theme originally penned by Gutenberg himself. They were both artists. They were both engineers. Only the four hundred years that separated them kept their shared condition from view.

Why should the connection seem more feasible to us today? The answer is simple: speed. Technology used to advance in slower, more differentiated stages. The book reigned as the mass medium of choice for several centuries; newspapers had a couple of hundred years to innovate; even film ruled the roost for thirty years before the rapid-fire succession of radio, then television, then the personal computer. With each innovation, the gap that kept the past at bay grew shorter, more attenuated. This meant little in the centuries-long increments of the book or the newspaper—not to mention the millennial scale of the cave painter—but as the stages grew more abbreviated, they began to interrupt the life cycles of individual humans. Rousseau lived his entire life under the spell of the printing press. Freud was born during the telegraph's heyday, and he made it all the way to the first stirrings of TV. There is a kind of knowledge to be found in those interruptions, those discontinuities, like the dialectical crosscuts of Eisenstein's *Battleship Potemkin*. (Yet another distinguished artist-engineer.) The explosion of media types in the twentieth century makes it possible for the first time to grasp the relationship of form to content, medium to message, engineering to artistry. A world governed exclusively by one medium is a world governed by itself. You can't measure a medium's influence without something to compare it with.

This, in fact, turns out to be the largely unsung message of McLuhan's *Understanding Media*. In a book full of radical pronouncements, the most suggestive—and puzzling—assertion comes near the end:

> At no period in human culture have men understood the psychic mechanisms involved in invention and technology. Today it is the instant speed of electric information that, for the first time, permits easy recognition of the patterns and the formal contours of change and development. The entire world, past and present, now reveals itself to us like a growing plant in an enormously accelerated movie. Electric speed is synonymous with light and with the understanding of causes.

There are a hundred books to be written about that "enormously accelerated movie." This is only one of them. But it is worth pausing for a second to be clear about what McLuhan is saying here. What made it possible for him to write *Understanding Media*, what makes it possible to concoct slogans like the "the medium is the message" in the first place, is the sheer velocity with which technology now advances. We can grasp the way different media shape our habits of thought because we can see the progression, the change from one form to another. You're born into a world dominated by television, and then suddenly you find yourself acclimating to the new medium of the World Wide Web. The shift is startling, even thrilling, depending on your mindset—but however you respond to the new forms, their arrival has an illuminating force. If you live your entire life under the spell of television, the mental world you inherit from the TV—the supremacy of images over text, the passive consumption, a preference for live events over historical contemplation—seems like second nature to you. Only when another medium rolls into view does the television's influence become perceptible. When those paradigm shifts arrive only once every few centuries, you have to be a genuine visionary or a lunatic to see beyond the limits of the form. McLuhan, of course, was a little of both.

Technological change has been a lightning rod for all manner of cultural electricity over the past two centuries. Think of the original Luddites, or the back-to-nature rejections of consumer society in the sixties. McLuhan may have been the most profoundly apolitical thinker in the second half of the twentieth century, but his ruminations on the consciousness-raising powers of technological speed sound uncannily like Karl Marx's—particularly the later Marx of the second and third volumes of *Capital.* "An anarchy of permanent revolution," Marx said of industrial society, not altogether unapprovingly. Where McLuhan saw electric speed as synonymous with the "understanding of causes," Marx saw its industrial equivalent as a force propelling us toward a working-class uprising. The dizzying, incessant waves of technological change—and their secondary effects on social organization—would prove unsustainable, maddening. The capitalist system would invariably come to crisis, reveal itself to be the crazed dervish that it was. The rate of change made it possible to think *historically* about a culture that liked to think of itself as outside history. It was a way of seeing beyond the current regime, and in doing so, it implicitly raised the possibility of salvation. The faster capital spun out the innovations, Marx argued, the more intolerable it would seem to those living under that accelerating clock. The revolution wouldn't be televised, but it would come from the same edgy, relentless drive for novelty that brought us television in the first place.

Labor historians talk about the great correction Western capitalism made in "allowing" the growth of unions to prevent genuine civil disturbances, but there is another, more subtle correction at work in the twentieth century: capitalism transformed technological speed from a looming, exponential threat—like global warming or overpopulation—into a lifestyle decision, a hip sensibility. Change could be our friend, the ads and the politicians leered. Embrace the speed and the unknowingness of electric society and all your troubles will subside in the rush, in the great leap forward. This is a line that runs like a neon thread through the last hundred years: from the Italian Futurists, fashioning their poetry after race cars and hand grenades, to the glib soothsaying of the AT&T "You will" campaign. And somewhere in this mix, buried beneath all the avant-garde euphoria and marketing hype, lay McLuhan, half-cheerleader and half-agnostic, shocked into recognition by the high voltages of twentieth-century machines. Technological acceleration wouldn't necessarily bring us contentment, he argued, but it would bring understanding. That was the great legacy of "electric speed."

The central topic of this book—the fusion of art and technology that we call interface design—is an offshoot of this same accelerated wisdom. We have reached a point where the various media evolve so rapidly that the inventors and the practitioners have blurred into one holistic unit, like a science lab hosting a creative-writing seminar. There are no artists working in the interface medium who are not, in one way or another, engineers as well. This has always been the case with culture and technology, of course; it's just that we used to pretend it was otherwise, by dutifully keeping the painters and the mechanics separate, on the college campuses, in the museum halls, on the bookshelves—wherever the twain had the slightest chance of meeting. The artisans of interface culture don't bother wasting time with these arbitrary divisions. Their medium reinvents itself too quickly for false oppositions between creative types and programmers. They have become something else, some new fusion of artist and engineer—interfacers, cyberpunks, Web masters—charged with the epic task of representing our digital machines, making sense of information in its raw form.

There is a kind of secret history to that fusion: Balzac was a printer, and his novels are obsessed with Gutenberg's technology. (His greatest novel, *Lost Illusions,* involves an ambitious tinkerer who concocts a new way to make paper stock.) The first motion-picture stars were the blushing, ill-at-ease relatives of the technicians behind the camera. But for the most part, we have kept the novelists and the mechanics, the painters and the programmers at separate ends of the spectrum, like two boxers restrained by a crowd of hangers-on and referees. Both camps have labored mightily to preserve a safe distance between the two, but the partition will not last long. We are due for a rumble.

A few final observations, and warnings, about the pages that follow. The first should be a comfort to readers who have tired of the recent bombast emanating from both the digital elite and their neo-Luddite critics. I have tried to keep this book as free of dogma and polemic as possible, emphasizing both the tremendous intellectual liberation of the modern interface and the darker, more sinister implications of that same technology. From its outset this book has been conceived as a kind of secular response to the twin religions of techno-boosterism and techno-phobia. On the most elemental level, I see it as a book of connections, a book of links—one in which desktop metaphors cohabit with Gothic cathedrals, and hypertext links rub shoulders with Victorian novels. Like the illuminations of McLuhan's electric speed, the commingling of traditional culture and its digital descendants should be seen as a cause for celebration, and not outrage. This is likely to trouble extremists on both sides of the spectrum. The neo-Luddites want you to imagine the computer as a betrayal of the book's slower, more concentrated intelligence; the techno-utopians want you to renounce your ties to the fixed limits of traditional media. Both sides are selling a revolution—it's just that they can't agree on whether it's a good thing. This book is about the continuities more than the radical breaks, the legacies more than the disavowals.

For that reason, the most controversial thing about this book may be the case it makes for its own existence. This book is both an argument for a new type of criticism and a working example of that criticism going about its business. For that reason alone, it may strike some readers as misguided. Throughout, I have tried to think about the elements of modern interface design as though they were the cultural equivalents of a Dickens novel, a Welles film, a Rem Koolhaas building—in other words, as works possessing great creative and social import, and having longer-term historical significance than just the latest product review in the high-tech trades. In the first section of the book, "Bitmapping," I discuss the origins of contemporary information-space and take a look at the way recent television programming anticipates the data filters of the present day. Each of the next five chapters focuses on one component of modern interface, exploring both the future possibilities of the device and its ties to the ancien régime of analog culture. "The Desktop" begins with the discovery of the office metaphor and then examines the difficulties of representing social life within that limited frame. "Windows" looks at the way multiple viewpoints change not only our psychological profiles but also our ethical and legal expectations about the proper use of information. "Links" draws an extended parallel between the hypertext medium of the World Wide Web and the grand synthetic narratives of the Victorian novel. The "Text" chapter makes a case for old-fashioned words on a screen, and explains how a computer managed to become a Shakespearean scholar. "Agents" takes a hard look at so-called intelligent software and speculates on how future interfaces may transform our cultural appetites. In the concluding chapter, "Infinity Imagined," I outline some of the broader themes that will hold sway over the new field of interface criticism in the next decade.

       I have tried to strike a balance in the following pages among technical explanations, historical narratives, and cultural analogies. Each chapter braids these three threads together, and I trust that the reader will find that stitch more enlightening than erratic. There are some wonderful stories behind the triumph of the graphic interface, and I have tried to convey a little of the texture and vitality of those events without turning this book into a historical account. I have also tried to keep the technical descriptions at a level that will appeal to power users and novices alike. My hope is that even the digitally savvy reader will find some revelation in the descriptions of more familiar objects, since part of a cultural critic's role is to make us think twice about experiences that are second nature to us. Of course, writing about technology from a cultural perspective invariably changes the subject matter in all sorts of surprising ways. Readers familiar with the contemporary high-tech landscape will note that I often discuss technologies that were not commercial successes and ignore some products that sold millions of copies. That is the unavoidable consequence of writing a book concerned more with imaginative breakthroughs than with box-office successes, and it will be familiar to anyone who follows the literary world, where mass appeal and aesthetic achievement rarely coexist in the same book. Fortunately, the history of the modern interface includes a number of instances in which mass audiences and creative innovation have managed to exist side-by-side. These happy coincidences of art and commerce are rarities in the cultural record: one thinks of Dickens in the Victorian age, Hitchcock's films in the fifties, the Beatles' recording career after *Rubber Soul*. The modern interface has had moments of comparable dexterity: a mass form that also labors at the cutting edge, a pathbreaker that still manages to attract an audience of millions. These intersection points are like the great eclipses of modern cultural experience, a rare and momentous alignment of forces, one that we may not see again for many years. We do well to take these alignments seriously anytime we are lucky enough to stumble across them. What follows is an attempt to do just that.

New York City

April 1997

BITMAPPING:
AN INTRODUCTION

In the fall of 1968 an unprepossessing middle-aged man named Doug Engelbart stood before a motley crowd of mathematicians, hobbyists, and borderline hippies in the San Francisco Civic Auditorium, and gave a product demonstration that changed the course of history.

It was an unlikely setting for such momentousness. The crowd brought to mind a Star Trek convention, or the wonderfully shabby trade show of private detectives and "security experts" from Coppola's *The Conversation*. Engelbart himself hardly conjured up images of Luther pounding out reforms on the church doors. But historians a hundred years from now will probably accord it the same weight and significance we now bestow on Ben Franklin's kite-flying experiments or Alexander Graham Bell's accidental phone conversation with Watson. Engelbart's thirty-minute demo was our first public glimpse of information-space, and we are still living in its shadow.

The idea of information-space had been around for thousands of years, but until Engelbart's demo, it was mostly just that: an idea. But what an idea it was! The Greek poet Simonides, born six centuries before Christ, was famous for his uncanny ability to build what rhetoricians call "memory palaces." These were the original information-spaces: stories turned into architecture, abstract concepts transformed into expansive—and meticulously decorated—imaginary houses. Simonides' trick relied on a quirk of the human mind: our visual memory is much more durable than our textual memory. That's why we're much more likely to forget a name than a face, and why we remember months later that a certain quote appeared on the upper-left-hand corner of a page, even if we've forgotten the wording of the quote itself. By imagining his stories as buildings, Simonides tapped that potential for spatial mnemonics. Each room triggered another event in the story, another twist in the argument. He could furnish the rooms for added detail, if he needed to stock up on adjectives or stylistic ornamentation. Telling the story itself was just a matter of strolling through the rooms of the palace.

The "memory palace" remained an essential tool in the art of rhetoric for thousands of years. Jonathan Spence's historical study *The Memory Palace of Matteo Ricci* tells the story of a seventeenth-century Italian missionary who attempts to convert China to Catholicism by teaching the natives a spatialized rendition of the Bible. Ricci's prodigious mnemonic skills were intended as both a sign of godliness (like the superior firepower or the pathogens of other missionaries) and the key to salvation. By entering into Ricci's sacred floor plan, and taking its passageways to heart, the Chinese too could enter the kingdom of heaven. Ricci's imagined landscapes were not uncommon for his time, and their influence spread far beyond the usual confines of academic rhetoric. As Spence notes, "The idea that memory systems were used to 'remember Heaven and Hell' can explain much of the iconography of Giotto's painting or the structure and detail of Dante's *Inferno,* and was common-place in scores of books in the sixteenth century."

There is a marvelous symmetry to this story: the venerable art of the memory palace, having aided the original "symbolic analysts" and "knowledge workers," returns to subdue the rhetorical complexity of the modern digital computer. In Engelbart's day, of course, computers weren't terribly skilled in the art of representation: the lingua franca of modern computing had been a bewildering, obscure mix of binary code and abbreviated commands, data loaded in clumsily with punch cards, and output to typewritten pages. A few pathbreakers like Ivan Sutherland had experimented with graphic displays, generating rudimentary polygons on blocky, pixelated screens. Sutherland's program—called Sketchpad— was the precursor to design applications like MacPaint and Photoshop. These are certainly impressive descendants, but Sketchpad mainly addressed the question of how you got the computer to draw things on the screen, how you got the machine to move beyond simply displaying characters. It didn't tackle the more significant problem of translating *all* digital information into a visual language. That was Engelbart's great quest, one he had been chasing for nearly two decades.

The quest had begun with a short provocative essay entitled "As We May Think" that Engelbart stumbled across while waiting to be shipped back to the States at the end of World War II. Written by a high-ranking army scientist named Vannevar Bush, the essay described a theoretical information processor called the Memex that enabled a user to "thread through" massive repositories of data, almost like a modern-day Web surfer. (We will return to Bush's seminal essay in the "Links" chapter.) The image haunted Engelbart for decades, as he followed a desultory career through the fledgling computer industry. That legendary demonstration in San Francisco was the first working product that even approached the functionality of Bush's speculative Memex device. Doug Engelbart has had a remarkably varied and visionary career, but even for that one demonstration alone, he deserves his reputation as the father of the modern interface.

What exactly is an interface anyway? In its simplest sense, the word refers to software that shapes the interaction between user and computer. The interface serves as a kind of translator, mediating between the two parties, making one sensible to the other. In other words, the relationship governed by the interface is a *semantic* one, characterized by meaning and expression rather than physical force. Digital computers are "literary machines," as hypertext guru Ted Nelson calls them. They work with signs and symbols, although this language, in its most elemental form, is almost impossible to understand. A computer thinks— if thinking is the right word for it—in tiny pulses of electricity, representing either an "on" or an "off" state, a zero or a one. Humans think in words, concepts, images, sounds, associations. A computer that does nothing but manipulate sequences of zeros and ones is nothing but an exceptionally inefficient adding machine. For the magic of the digital revolution to take place, a computer must also *represent itself* to the user, in a language that the user understands.

In this sense, the term *computer* is something of a misnomer, since the real innovation here is not simply the capacity for numerical calculation (mechanical calculators, after all, predate the digital era by many years). The crucial technological breakthrough lies instead with this idea of the computer as a symbolic system, a machine that traffics in representations or signs rather than in the mechanical cause-and-effect of the cotton gin or the automobile. In this respect, computers have a kind of surface resemblance to older technologies—the Gutenberg press, say, or a Cinemascope camera. But there is an important distinction. A printing press or a camera deals with representations as end-products or results; these machines are representational in that they print words on paper or record images on film, but the underlying processes are purely mechanical in nature. A computer, on the other hand, is a symbolic system from the ground up. Those pulses of electricity are symbols that stand in for zeros and ones, which in turn represent simple mathematical instruction sets, which in turn represent words or images, spreadsheets or e-mail messages. The enormous power of the modern digital computer depends on this capacity for self-representation.

More often than not, this representation takes the form of a metaphor. A string of zeros and ones—itself a kind of language, though unintelligible to most humans—is replaced by a metaphor of a virtual folder residing on a virtual desktop. These metaphors are the core idiom of the contemporary graphic interface. As idioms go, they are relatively simple ones, which is why for most PC users, the idea of interface design as a legitimate art form will probably sound somewhat hyperbolic. The word *interface* itself conjures up cartoon images of colorful icons and animated trash cans, as well as the inevitable saccharine platitudes of "user-friendliness." That these associations should spring to mind so readily testifies to the extraordinary success of the graphical user interface (or GUI), first developed at Xerox's Palo Alto Research Center during the seventies and subsequently popularized by the Apple Macintosh. The widespread adoption of the GUI has dramatically changed the way in which humans and computers interact, and has greatly expanded computer literacy among people once alienated by the arcane syntax of the older "command line" interfaces. The visual metaphors that Doug Engelbart's demo first conjured up in the sixties probably had more to do with popularizing the digital revolution than any other software advance on record.

Dreaming up metaphors for new machines has, of course, a long and distinguished history. Every age comes to terms with the latest technology by drawing upon imagery of older and more familiar things. Usually this takes the form of an analogy between machines and organisms. Dickens saw the Manchester factories as mechanical jungles, populated by "serpents of smoke" and a steam engine with a "head like an elephant." Thoreau speculated darkly on "that devilish Iron Horse" cutting its way across the American landscape, while Thackeray imagined the British railway system as arteries coursing through the body politic. The term *computer* itself derives from low-tech roots: computers were human calculators in the days before digital code, workers skilled with the slide rule and old-fashioned long division.

It may be that every high-tech innovation is accompanied by imaginative flashbacks of this sort, but our own historical moment has added an unusual twist to this long tradition. Organic, low-tech metaphors once belonged to those lagging behind the machinic power curve, the Luddites and the antediluvians, the poets and the novelists, the ones reaching for older analogies because the shock of the new had so overwhelmed them. In today's society, the task of translation has migrated to the technicians. In the age of the graphic interface, with its visual metaphors of trash cans and desktop folders, imaginative flashbacks have become programming feats, conjured up by high-tech wizards hacking away in assembly language. Where the Victorian novel shaped our understanding of the new towns wrapped around the steel mill and the cotton gin, and fifties television served as an imaginative guide to the new suburban enclaves created by the automobile, the interface makes the teeming, invisible world of zeros and ones sensible to us. There are few creative acts in modern life more significant than this one, and few with such broad social consequences.

Some readers may find this claim overstated. However innovative the bitmaps of the graphic interface have turned out to be, it's still quite a leap from writing *Bleak House* to designing software that facilitates writing business proposals or drawing pie charts. The Macintosh interface may have made it easier to write the Great American Novel, but surely that doesn't imply that software and literature belong to equivalent cultural categories. Why make such sweeping claims for the interface?

This is where the rise of the Internet becomes important. The first generation of graphic interfaces (like the Mac or Windows) seem so incommensurate with our notions of "high art" because the tasks they represent on screen are relatively simple ones. In everyday use, a PC functions mostly as a glorified typewriter, or file cabinet, or calculator; effective interface design enables a single user to navigate intuitively through his or her documents and applications, occasionally communicating with the outside world via fax or e-mail. The simplicity of the interface reflects the simplicity of the tools offered by the computer itself. But in the past few years, new tools have appeared on the horizon, tools that will transform our basic assumptions about the computer and its broader social role. (In fact, "tool" wouldn't seem to be the word for it anymore, since what is now emerging is really more like an environment, or a space.) As our machines are increasingly jacked into global networks of information, it becomes more and more difficult to *imagine* the dataspace at our fingertips, to picture all that complexity in our mind's eye—the way city dwellers, in the sociologist Kevin Lynch's phrase, "cognitively map" their real-world environs.

Representing all that information is going to require a new visual language, as complex and meaningful as the great metropolitan narratives of the nineteenth-century novel. We can already see the first stirrings of this new form in recent interface designs that have moved beyond the two-dimensional desktop metaphor into more immersive digital environments: town squares, shopping malls, personal assistants, living rooms. As the infosphere continues its exponential growth, the metaphors used to describe it will also grow in both scale and complexity. The agora of the twenty-first century may very well relocate to cyberspace, but it won't get very far without interface architects to draw up the blueprints.

All of which raises another question: By what criteria should we judge our interfaces? If the interface medium is indeed headed toward the breadth and complexity of genuine art, then we are going to need a new language to describe it, a new critical vocabulary. Some of this language will rise up sui generis out of the new technologies, but most of it will borrow extensively from preexisting traditions: art and architecture, the cinema and the novel. Certain digital revolutionaries will see this pilfering from the past as a limitation, the telltale sign of a thinker still trapped in the analog world of the past. But the truth is, radical breakthroughs are anomalies in the cultural fossil record. The interplay between past and future forms drives the creative process more than it impedes it. Interface designers have much to learn from the invention of Renaissance perspective, or the buildings of Frank Gehry, and interface critics have much to learn from the interpretative schools that have developed around those older movements. We need a new language to describe the new medium of interface, but that doesn't mean we can't borrow some of our terminology from the forms that have come before it.

How should we understand the cultural import of interface design in today's world? Put simply, the importance of interface design revolves around this apparent paradox: we live in a society that is increasingly shaped by events in cyberspace, and yet cyberspace remains, for all practical purposes, invisible, outside our perceptual grasp. Our only access to this parallel universe of zeros and ones runs through the conduit of the computer interface, which means that the most dynamic and innovative region of the modern world reveals itself to us only through the anonymous middlemen of interface design. How we choose to imagine these new online communities is obviously a matter of great social and political significance. The Victorians had writers like Dickens to ease them through the technological revolutions of the industrial age, writers who built novelistic maps of the threatening new territory and the social relations it produced. Our guides to the virtual cities of the twenty-first century will perform a comparable service,

•

b
i
t
m
a
p
p
i
n
g

only this time the interface—and not the novel—will be their medium.

Doug Engelbart, of course, was one of the very first to grasp how essential information guides would become, and his vision of how such guides would work—the vision outlined in that 1968 demo—still defines the basic blueprint of the modern interface. Like most technological revolutions, Engelbart's information-space involved several key components, each of which contributed something essential to the larger whole. There was, first of all, the wonderful idea of bitmapping (technically refined by the savants at Xerox PARC in the subsequent years). The word itself suggested an unlikely alliance of cartography and binary code, an explorer's guide to the new frontier of information. Each pixel on the computer screen was assigned to a small chunk of the computer's memory: on a simple, black-and-white screen, that chunk would be a single bit, either a zero or a one. If the pixel was lit up, the value of the bit would be "one"; if the pixel went dark, its value was "zero." The computer, in other words, imagined the screen as a grid of pixels, a two-dimensional space. Data, for the first time, would have a physical location—or rather, a physical location *and* a virtual location: the electrons shuttling through the processor, and their mirrored image on the screen.

But once you endowed that data with spatial attributes, what were you going to do with it? Engelbart's great breakthrough involved the principle of *direct manipulation*. It was one thing to represent a text document as a window or an icon, but unless the user had some control over those images, the illusion would be remote, unconvincing, like a film projected at only a few frames per second. For the illusion of information-space to work, you had to be able to get your hands dirty, move things around, make things happen. That's where direct manipulation came in. Instead of typing in obscure commands, the user could simply point at something and expand its contents, or drag it across the screen. Instead of telling the computer to execute a particular task—"open this file"—users appeared to do it themselves. There was a strangely paradoxical quality to direct manipulation: in reality, the graphic interface had added another layer separating the user from his or her information. But the tactile immediacy of the illusion made it seem as though the information was now closer at hand, rather than farther away. You felt as though you were doing something directly with your data, rather than telling the computer to do it for you.

Of course, you needed a good tool to do all that newfound handiwork. Engelbart came up with two. The first was an ingenious replacement for the QWERTY keyboard that used a "chording" system of keystrokes, where each symbol was represented by several keys pressed simultaneously. It was markedly faster than a traditional keyboard, particularly when used with software that had been optimized for it. Unfortunately, it required that you learn how to type all over again, a demand that proved too strenuous for the device ever to attract a mass audience. But the other input tool that Engelbart used that fall afternoon in San Francisco did eventually develop a market, though it would take more than a decade for it to take shape. Engelbart called his device the "mouse."

As in the current incarnation, Engelbart's mouse served as the user's representative in dataspace. The software coordinated between the user's hand movements and a pointer on the screen, allowing Engelbart to click on windows or icons, open and close things, rearrange the information-space on the monitor. The pointer darting across the screen was the user's virtual doppelgänger. The visual feedback gave the experience its immediacy, its directness: move the mouse an inch or two to the right and the onscreen pointer would do the same. Without that direct link, the whole experience would have been more like watching television, zoning out beneath a steady stream of images that are kept separate from you, distinct. The mouse allowed the user to enter that world and truly manipulate things inside it, and for that reason it was much more than just a pointing device.

With its seamless integration of bitmapped info-space, direct manipulation, and the mouse, Engelbart's show electrified the crowd. They had seen nothing like it before, and many of them would wait years to see its equivalent again. The luminous new world of information-space had suddenly come into view, and it was clear that the future of computing had been changed irrevocably. Howard Rheingold described the revelation best in *Tools for Thought:*

> The territory you see through the augmented window in your new vehicle is not the normal landscape of plains and trees and oceans, but an informationscape in which the features are words, numbers, graphs, images, concepts, paragraphs, arguments, relationships, formulas, diagrams, proofs, bodies of literature and schools of criticism. The effect is dizzying at first. In Doug's words, all of our old habits of organizing information are "blasted open" by exposure to a

system modeled, not on pencils and printing presses, but on the way the human mind processes information.

That informationscape was both a technological advance and a work of profound creativity. It changed the way we use our machines, but it also changed the way we imagine them. For centuries, Western culture had fantasized about its technology in *prosthetic* terms, as a supplement to the body, like a wooden leg or a telescope. The great industrialists rhapsodized about the cotton gin as a miraculous extension of the hand-weaver's fingers, while critics like Dickens and Zola thundered against the indecency of turning machinists into so many machines. This tradition continued well into the twentieth century. The French novelist Céline captured the violent hybrids of a Ford automobile plant circa 1935 in this famous passage from *Journey to the End of Night:*

> Everything trembled in the enormous building, and we ourselves, from our ears to the soles of our feet, were gathered into this trembling, which came from the windows, the floor, and all the clanking metal, tremors that shook the whole building from top to bottom. We ourselves became machines, our flesh trembled in the furious din, it gripped us around our heads and in our bowels and rose up to the eyes in quick continuous jolts.

Of course, the machine-as-prosthesis imagery was not always so bleak. The Futurist poet Marinetti enthused over the new hybrids of "man-torpedo-boat" and even elevated these mongrels into a poetic call to arms, delivered most forcefully in the last lines of *The Futurist Cookbook:* "Through intuition we will conquer the seemingly unconquerable hostility that separates our human flesh from the metal of motors." McLuhan himself never tired of describing the electric technologies of the twentieth century as extensions of our central nervous system. Even Vannevar Bush explicitly positioned his Memex device as a tool for "augmenting" our intellect, the way a pair of spectacles might augment our vision.

Perhaps as an implicit homage to his long-distance mentor, Doug Engelbart used the language of "augmentation" throughout his career, but the bitmapped datasphere he unleashed on the world in 1968 was the first major break from the machine-as-prosthesis worldview. For the first time, a machine was imagined not as an attachment to our bodies, but as an environment, a space to be explored. You could project yourself into this world, lose your bearings, stumble across things. It was more like a landscape than a machine, a "city of bits," as MIT's William Mitchell called it in his 1995 book. Not since the Renaissance artisans hit upon the mathematics of painted perspective has technology so dramatically transformed the spatial imagination. Most of today's high-tech vocabulary derives from this initial breakthrough: cyberspace, surfing, navigating, webs, desktops, windows, dragging, dropping, point-and-clicking. The idiom begins and ends with information-space. And that's only a few decades after Engelbart's original demo—imagine how far the metaphor will have traveled by the end of the next century.

We are all bitmappers now, thanks to Doug Engelbart's information-space. The poets and the inventors of the past few centuries rigged up our machines as extensions of our bodies, augmentations, supplements. But Engelbart's information-space "blasted open" that tradition, making room for the modern interface in the process. The industrial age gave us our prosthetic limbs and man-torpedo hybrids, but Doug Engelbart gave us the first machine worth living in. We are only now beginning to understand the magnitude of that gift.

Imaginative transformations of this scope do not take place in a vacuum. They are invariably accompanied by more lateral effects, unintended consequences that ripple out into other fields. The automobile changed the way our cities developed, pushing our old urban cores outward toward the satellite or "edge" cities of today. The psychological introspection of the late-nineteenth-century novel paved the way for psycho-analysis, which in turn made possible the self-help and pop-psychology movements of the postwar age. The interface has already changed the way we use our computers, and it will con-tinue to do so in the years to come. But it is also bound to change other realms of modern experience in more unlikely, unpredictable ways. This book is in part an account of these strange migrations, the new medium of interface design winding its way through a broad swath of modern life, some-times far removed from the computer screen.

Perhaps the most vivid example of this lateral movement can be seen in the endlessly self-referential loop of nineties television programming. As the traditional narra-tives—the sitcoms and soap operas and talk shows—plod reli-ably through their conventional paces, the *meta*shows, the shows interested not in telling stories but in riffing on other media, have experienced a genuine flowering in recent years—

not just in cultural significance but also in genetic diversity. These are the TV shows and video installations devoted exclusively to "reading" the media, and not that old fuddy-duddy practice of storytelling. These are the real innovators of modern television, attaching themselves like parasites to larger, more sluggish host organisms, and replicating indirectly through them. On the surface, they appear to belong to a wholly different genus from the bitmaps of interface design. And yet the parasites of nineties' TV turn out to be intimately related to Doug Engelbart's information-space, though to make sense of the connection, we have to zoom out a little, far enough to perceive the larger shape of the metaform itself.

Forty years ago you could count the parasite forms on your fingers: *Reader's Digest, TV Guide,* the fanzines devoted to assorted Hollywood stars and doo-wop heartthrobs. Everything else dealt with stories about folks somewhere outside the fun house of mass media, even as that old, unmediated realm became increasingly remote, until the thought of stories about lives untouched by the mass media became for all practical purposes unthinkable, the way stories that involved only knights and princes became unthinkable in the eighteenth century. The technological changes that ushered in merchant capitalism did away with the old, aristocratic morality plays and introduced a new, rougher form—the realist novel, with its orphans and scoundrels and wayward heroines. In the same way, the electric technologies of the twentieth century have done away with the old storytelling forms, or at least scaled them down to assembly-line repetition, while simultaneously unleashing a flock of new organisms into the larger cultural ecology.

But here's the rub: these new organisms don't tell stories. They riff, annotate, dismantle, dissect, sample. Everything they do refracts back onto some other "straight" media, on which they rely for their livelihood. They relate to their story-driven predecessors the way a movie review relates to a movie. But unlike your everyday film critic, these new forms run the gamut from high to low culture, from mass appeal to indie cachet. Think of the variations unleashed in the past ten years alone: MTV's *Zoo TV,* Paper Tiger Television, the E Channel, Comedy Central's *Daily News,* VH-1's *Pop-Up Video.* Half the shows on public-access cable here in Manhattan consist of inveterate channel surfers introducing their "clips of the week," as though what you watched on TV over the past few days now passed for self-expression. The tabloid "news" shows—*Hard Copy, Extra, A Current Affair*—regularly manage to make news out of pure mediation: Mick Jagger videotaped stumbling out of the Viper Room, Ivana Trump captured strolling down Worth Avenue. In the old days you at least had to hold a press conference to concoct a pseudo-event; now simply being caught on camera suffices.

There may not be a great deal of "quality programming" in this mix, but the sheer quantity of this new genre—the diversity of the species—is remarkable. All the evidence suggests that the metaforms are evolving at a much faster clip than their storytelling competitors. The sitcom has learned a few new tricks in the past decade—Seinfeld's "show about nothing," the surreal landscape of *The Simpsons*—but most of the network comedy programming could have easily aired fifteen years ago, give or take a few wardrobe changes. Even if you factor in the lesbian plot twist, the basic structure of an *Ellen* episode looks a lot like an average *Taxi* from 1981, while a few nights with *The Honeymooners* makes *Roseanne*—herself considered something of a pathbreaker—look like a poor relation living shamelessly off the family name. But while the sitcom has languished, the self-referential "commentary" show has come into its own. Twenty years ago, the "meta" genre didn't even exist on TV. What, for instance, was the seventies' equivalent of *Beavis and Butt-head?* Or CNN's navel-gazing *Reliable Sources?* Or Comedy Central's *Mystery Science Theater?* Or the E Channel's *Talk Soup?* Or, for that matter, *anything* on the E Channel?

Television storytelling—as a medium—has made only modest advances over the past twenty years, which is why *Melrose Place* looks so much like *Dynasty,* and why Archie Bunker still seems raw-edged and scandalous a quarter of a century after the fact. But televised "riffing"—television commenting on television—has undergone a period of dramatic growth. Most critics who have addressed the growth of the "meta" form have acted as though there were something abnormal, something malignant about it. They talk about the phenomenon using the language of epidemics, not unlike the way the William Bennetts and David Denbys of the world talk about the epidemic of violence on television. The surge in shows about shows is felt to be pathological, a symptom of something rotten lurking in the mediasphere Denmark.

There is something illusory about this contempt for the mass-media illusion. You can see this most clearly in those critics who endlessly replay the Nixon-Kennedy debates the way other, more conspiratorial critics cling to the Zapruder film. (If only Nixon had worn makeup!) To be sure, the descent-into-media-purgatory narrative has its merits. No one seriously defends the deep mediation of the contemporary electoral process, with its image makers and spin doctors and sound bites, and its candidates selected more for their skill with an open mike than their legislative record. But modern imagineering should provoke more than just empty nostalgia for a kinder, less-mediated America. For the older critics of image society, the false god of television is cause for endless hand-wringing, for long, contemplative essays on the decline of the stump speech and the kissing of babies. For the new parasite forms, however, the society of the spectacle is a call to arms. Instead of muttering "the emperor has no clothes" from the sidelines, they've stripped naked and jumped aboard the mass-media float. What better way to call the emperor's bluff?

It's worth stressing here that the shift from story-telling to commentary—from host organism to parasite—is more than just standard-issue postmodernism. The television shows that have gravitated toward metacommentary in the past few years have done so not because they have given up on "the real," as the French psychoanalysts like to say. They have attached themselves to the mass-media body because the mass media is now a fundamental, irreversible component of their everyday life, as inescapable as all the old inescapables—sex, death, taxes, you name it. The infosphere is now a part of our "real life"—which makes commenting on it as natural as commenting on the weather. The older tradition of media criticism—Daniel Boorstin's classic work *The Image* being the ultimate example—sees the tendency for self-reference as a kind of hall-of-mirrors effect, where the real body politics of face-to-face existence slouch toward a vanishing point of end-less reflection. As Doug Rushkoff puts it in *Media Virus:* "Philosophers who grew up before television ... view media or even technology, for that matter, as something outside the realm of the natural. To them, media can only display or comment on something real. They cannot acknowledge that the media is something real itself, something that exists on its own and that might have its own needs and agendas."

The older critics of image society suffer from a vampire complex. They're in the business of renouncing mirrors. But for most people raised on television—and particularly those of us raised on the idle promenades of cable-era surfing—forswearing the mass-media mirror is like forswearing gravity. (Sure, you'd like to do it on occasion, but the rules are the rules.) The growth of the parasite form reflects the increasingly naturalized role of the media in contemporary life. We are fixated with the image not because we have lost faith in reality, but because images now have an enormous impact on reality, to the extent that the older image–reality opposition doesn't really work anymore. This pattern of renunciation and acceptance has a long history. New technologies invariably possess the aura of unreality at their outset, and then march steadily toward the natural world. Most commentators on the new steam-driven factories of the early nineteenth century—Dickens and Carlyle in England, Melville in the United States—were as struck by the monstrous artificiality of the machines as by the squalor of the working-class conditions that surrounded them. (The onslaught of noise alone generated hundreds of descriptive passages; it was an experience as foreign to the Victorians as weightlessness would be to us today.) But the strangeness, the unreality eventually wears away. By the dawning of the information age, that artifice had become second nature in most of the West, which is why the daily grind of assembly-line existence is felt to be more "real" than the illusory reveries of television.

Just as twentieth-century labor movements grounded themselves in the authenticity of industrial work, the new parasite forms take the images of contemporary television at face value—as a real, vital, unavoidable component of everyday life. It just happens that this particular region of everyday life is not well represented by *stories*. The reality of nineteenth-century society—the "manifold awakenings of men to endurance and labor," as George Eliot famously described it in *Middlemarch*—required the steady, deep-focus narratives of the triple-decker novel. Eliot's "involutary, palpitating life" is alive and well in the mediated universe of late-twentieth-century culture. Only the palpitations of modern life don't readily map onto the sprawling plotlines of the Victorian novel, or even the "high-concept," cookie-cutter narratives of Hollywood films. Instead of stories, we have riffs, annotations, asides. We are a nation of commentators, either heckling the mass-media magnates from the cheap seats of our living rooms or cheering on the hecklers that are delivered to us there: Beavis and Butthead, the political pundits and media watchers, the robots of *Mystery Science Theater.*

It is customary to see the metaforms as an offshoot of the culture's congenital case of irony—what used to be called disaffection, back before irony developed its own radical politics. Gen X critics like Rushkoff imagine slacker detachment as a kind of armchair subversion, deconstructing the house of mass media from within. Irony does of course run deep in many of the new parasite forms, but there are plenty of specimens in the fossil record where the tonal DNA is markedly less derisive. However bewitching the idea of a living-room liberation front may be, it *is* hard to imagine *Beavis and Butt-head* as a platform for co-opting the mass-media apparatus. In fact, you can easily interpret the show as the ultimate in couch-potato passivity: instead of composing our own wisecracks about the Fugees video or the infomercial or the rerun, we now outsource them to a couple of animated malcontents on the screen. So much for interactivity!

No, the defining characteristic of the parasite form is neither hall-of-mirrors delusion nor dissident irony. What unites the diverse strains of this emergent species is a shared belief in the need for information filters—data making sense of other data. The parasite forms thrive in situations where the available information greatly exceeds our capacity to process it. Metaforms prosper at those threshold points where the signals degenerate into noise, where the datasphere becomes too wild and overwrought to navigate alone. In these climates, all manner of metaforms appear: condensers, satirists, interpreters, samplers, translators. They feed on surplus information, on the bewildering sensory overload of the contemporary mediasphere. And this is where the connection to the modern interface comes into focus.

Where the novel ushered its readers through the crowds and assembly lines of industrial life, the metaforms process and contextualize the byzantine new reality of information overload. They serve as buffers, translators, tour guides. Unlike the novel, they prefer evaluation and interpretation to storytelling and character development. The old-style narratives acculturated their audience to the industrial age by building elaborate structures of cause and effect, connecting the increasingly atomized public spaces of the new cities, linking working-class orphans to withered aristocrats to idle speculators to colonial scavengers. These narrative webs— dense and meticulously interwoven—were a way of restoring a sense of connection, of unity, to a culture that had transformed itself utterly in the space of fifty years. The novel was the answer to the question: "What connects all these bewildering new social realities?" And that answer was phrased in the form of a story. The parasite forms, on the other hand, are a response to the question: "What does all this information *mean?* Which sources are the most reliable ones? How does this information relate to my own particular worldview?" That response arrives as a kind of hybrid, a mix of metaphor, footnote, translation, and parody. It is a measure of the newness of the form that we lack a single word to describe it.

Here, of course, we come up against the issue of aesthetic taste and distinction, and the criteria we use to evaluate these programs. Even if you accept the analogy between the industrial-era novel and the metaforms of the information age, surely there remains a qualitative distinction to be drawn between *Great Expectations* and *Mystery Science Theater,* between *Germinal* and *Talk Soup*. Both forms arise out of the turbulence of their respective periods, and both offer a symbolic corrective or solution to that turbulence, a sort of cognitive Dramamine. But those solutions represent two very different orders of achievement. The blurring of highbrow and lowbrow can only go so far, after all. A cultural criticism that can't recognize the distinction between the social complexity of *Middlemarch* and the wiseass schlock of *Mystery Science Theater* is just as negligent and shortsighted as a criticism that only values the work of Dead White European Males. Surely there's a way to talk about pulp TV and the novel as analogs of each other without admitting *Beavis and Butt-head* into the canon of Great Works. It's true enough that Shakespeare was the MTV of his day—as pop aficionados like to remind us—but that doesn't necessarily mean the reverse is also true. (Would that make VH-1 our Ben Jonson?) Anyone who seriously thinks that MTV is the Shakespeare of our time would probably do well to have the cable cut off for a few months, just to get some perspective on it all.

Despite the preachings of our pop Cassandras, the general mediocrity of the parasite forms is not a sign of the decline of Western civilization, or the closing of the American mind. The shows may seem for the most part fatuous and one-dimensional, just a few steps up from a drunk heckling a stage show, but those limitations are not to be blamed on some conspiracy to "dumb down" our mass entertainment. If our metaforms do turn out to be underachievers by the usual cultural yardsticks—and all the early returns indicate that they will—this deficiency suggests a more interesting pattern, one that comes close to functioning as a general law in the evolution of media types. The metaforms seem so disappointing because they are taking on a symbolic task that exceeds the capacity of their medium. The new parasites remain parasites because they are, in a word, too *hot* for their environment. They float across our television screens as hints and intimations, a glimpse of the future shrouded in the worn, restrictive garments of the past, like a Cubist body rigged together with corsets and lace. They are ghosts of technologies to come.

This, too, is an old story, in its way. At major transition points, where one platform or genre gives way to another, the older form invariably strains to approximate the rhythms and mannerisms of the emergent form. There is something primal, almost irresistible about this pull, like a field of sunflowers leaning toward the light of a new day. The older medium wants to reinvent itself—chrysalis-style—in the image of the new, but its existing conventions won't allow such a dramatic transformation. This is the story of Satan in *Paradise Lost*—the dashing, bewitching rogue at the center of

Milton's sacred poem. The form itself belongs very much to its period—a Christian epic, loaded with allusions to religious tumult in Restoration England. But the character of Satan himself belongs to the future. Writing nearly two centuries later, Percy Shelley described Milton's Satan as the original "arch-Rebel," the first in a long cultural history of bad guys who swept readers off their feet. There's no room for such a presence in a work governed by a rigid moral universe, which is why Satan seems so out of place to us now in Milton's blank verse. But the character soon found another home, in another genre altogether: the novels of the nineteenth century, with their morally ambiguous protagonists, their charlatans and con artists and ill-mannered young men. Milton's Lucifer is the long-lost ancestor of Stendhal's Julien Sorel or Flaubert's Frederic Moreau: devious and corrupt, sure, but also passionate, even sexy. If Hamlet inaugurates the contemporary malaise of paralyzed introspection, Satan lies at the source of another modern condition—the allure of the illicit. Only Milton's Satan was conceived before his time, before the literary world had produced a mature form that could do justice to the character, and not struggle against it.

The principle at work here is this: at transition points, some messages may evolve faster than their medium. And in doing so, they anticipate another medium, one that is still in embryo. The history of radio follows the same pattern. For most of the thirties and forties, as television slowly migrated from the R&D lab to the RCA showroom, radio programming experimented extensively with a format that was particularly ill-suited to the medium: "radio theater," theatrical narratives relayed without faces, sets, costumes, action—just voices, background music, the occasional sound effect, and a word or two from the sponsor. Given the visual inventiveness and expertise of Hollywood directors from that period (Welles, Capra, Huston), it's amazing that anyone even listened to these technologically challenged productions, much less dubbed them the "golden age" of radio. But with hindsight, we can see that classic shows like *The Shadow* and *The Jack Benny Show* weren't really great *radio* programs—they were just bad *television* shows, TV-style narratives stripped down to fit the limited dimensions of radio. They were a message waiting for their medium to come. Old-school Sid Caesar fans may object to this appraisal, but consider the evolution of radio after television took off in the fifties: the narrative shows maintained their audience for a decade or so, and a few anomalies carried on after that (I remember listening to *Mystery Theater* with E. G. Marshall in the late seventies). But the great bulk of radio programming gravitated steadily toward content that actually suited the platform—music, news, and talk—and left the audio dramas for the books-on-tape crowd.

We are at another such intersection point today. The parasite forms—the talk shows about talk shows, the info-guides and video activists, the animated hecklers and media critics—make up a kind of advance guard, a television pilot for a series that will run on another medium. They are all straining to do something within the TV box that cannot be done, for fundamental technological reasons. The parasite forms are finally all about meta-information—or better yet, *flexible* meta-information. For the first time in its history, television has begun to offers its viewers different lenses through which to view the "real" content, which turns out to be another television show. The raw data can now be consumed through different mediations. You can watch the music video directly on MTV, of course, but you can also watch it through the filter of Beavis and Butt-head, with their staccato comments running in the background. Or you can watch the same video on MTV's *Yak Live,* with a stream of real-time comments rolling beneath the image, piped in directly from an AOL chat room, like a crossbreed of liner notes and bathroom graffiti. *Beavis and Butt-head* and *Yak* are metaforms, filters; you still take in the video itself, but the experience is necessarily transformed (if not always enhanced) by the filter that conveys it to you.

Needless to say, these filters are not particularly sophisticated ones. The fact that these new lenses, these mirrors within mirrors, have sprung into existence with such speed shouldn't distract us from the basic fact that they have their limits. Those limits stem directly from the technological restrictions of the television box itself. The information that arrives on your TV screen is hardwired and inflexible; you consume what the networks tell you to consume. Your only feedback mechanism is the remote control. (Most "information filtering" is still of the changing-channel sort.) The metamedia that does manage to slither onto the screen is just a hack job, like rigging up a toaster oven to control temperature variation at a power plant. You *can* create an information filter on the boob tube, but you're working with damaged goods all the while. The whole motley crew of parasite forms on the screen today brings to mind Dr. Johnson's remark about the dog walking on its hind legs: it is not done well, but one is surprised to see it done at all.

Twenty or thirty years from now we will see the outbreak of televised parasites as a kind of evolutionary oddity, a distant ancestor that shared a few strands of cultural DNA with the modern species but that never thrived in its own ecosystem. Metaforms don't fare well in the analog world of television, where the signal is only as malleable as the hue and saturation knobs on your set. But the digital world is another story. That world—the rich, expansive frontier of PCs, ATMs, videodiscs, World Wide Webs, personal communicators, intelligent agents—is the home planet of information filters. The parasite forms are a fringe benefit on analog TV, a flourish. In the digital world, they are a fact of life. There is no such thing as digital information *without* filters, for reasons that will become increasingly clear. As more and more of the culture translates itself into the digital language of zeros and ones, these filters will become enormously important, even as their cultural roles become increasingly diverse, embracing entertainment, politics, journalism, education, and more. What follows is an attempt to see all these various developments as instances of a larger idea, a new cultural form hovering somewhere between medium and message, a metaform that lives in the nether land between information producer and consumer. The interface is a way of mapping that strange new territory, a way for us to get our bearings in a bewildering environment. Decades ago, Doug Engelbart and a few other visionaries recognized that the information explosion could be both destructive *and* liberating—and without a metaform to guide us through that information-space, we'd run the risk of losing ourselves in the surplus information. It is a testament to the power and radiance of this idea that television itself has adopted some of its basic values over the past few years, albeit in the clumsy, two-dimensional form of *Mystery Science Theater* and *Talk Soup*.

The rest of this book will concern itself with the fate of this metaform in the *digital* world, since that is its natural habitat.

But the parasite's mixed success in the analog soup of late-twentieth-century television drives home an important point, one that will be a recurring theme of this book. No significant cultural form springs into existence fully realized. There is always a gestation period, where the divisions between different genres, conventions, or media types are less defined. These transition points can be disorienting to the societies living through them, and some of that disorientation is of a taxonomic variety, the confusion of creating categories for—and perceiving relationships between—things that are not easily categorized.

The preceding pages may appear to have played into that confusion. On the face of it, *Talk Soup* and a bank ATM don't have much in common. Why yoke Butt-head to the World Wide Web, when he so clearly belongs to television? The answer is that cultural forms are not always reducible to the physical medium that sustains them. Often the bedfellows are so strange that their dalliances don't become visible for decades. Imagine time-traveling back to Moscow circa 1924 and informing Sergei Eisenstein that his highbrow innovation of cinematic montage would utterly transform the American pop music industry via the rapid-fire, retina-overload editing of most MTV videos. The history of cultural innovation is littered with these unlikely alliances. The "stream-of-consciousness" device pioneered by Joyce in *Ulysses* grew out of the mono-logues of Shakespearean drama, the psychological introspection of James and Freud, and the new impulse-oriented "science" of advertising. (Joyce cast an advertising agent, Leopold Bloom, as the hero of his modern-day Odyssey, even though the profession was still in its infancy.) Contending that the stream of consciousness "belongs" to the novelistic medium makes sense in a heavily specialized academic environment, but it does a terrible injustice to the real history of the device.

We like to imagine that our cultural archetypes have finer pedigrees than they do. But the truth is most of them are mongrels. This should be a reason for celebration. The cross-breeding between different species is one of the great advantages that cultural evolution has over the old-fashioned Darwinian variety. As Stephen Jay Gould writes in *Full House,* "[Biological] species do not amalgamate or join with others. Species interact in a rich variety of ecological ways, but they cannot physically join into a single reproductive unit.... Cultural change, on the other hand, receives a powerful boost from [the] amalgamation and anastomosis of different traditions." In the cultural sphere, the hybrids are stronger, more innovative, more robust than the pure breeds. And that is why there is still value to be found in the parasite forms of analog television. Like *Middlemarch's* Dorothea Brooke, these forms are "the mixed result of a young and noble impulse struggling amidst the conditions of an imperfect ... state." They are digital forms trapped in an analog medium. As such they remind us again of the significance of the shift from analog to digital, a shift that is as much cultural and imaginative as it is technological and economic. The digital revolution will surely transform stock markets and library research and credit profiles, as the pundits on the business pages have been predicting for years. But it also promises to transform our experience of the world, just as the Industrial Revolution transformed the experiences of nineteenth-century Westerners. This book is in part an attempt to imagine the nature of this new experience, to sketch out its properties in advance.

In the early sixties McLuhan famously remarked that living with electric and mechanical technologies at the same time was "the peculiar drama of the twentieth century." The great drama of the next few decades will unfold under the crossed stars of the analog and the digital. Like the chorus of Greek tragedy, information filters will guide us through this transition, translating the zeros and ones of digital language into the more familiar, analog images of everyday life. These metaforms, these bitmappings will come to occupy nearly every facet of modern society: work, play, romance, family, high art, pop culture, politics. But the form itself will be the same, despite its many guises, laboring away in that strange new zone between medium and message. That zone is what we call the interface.

t w o

THE DESKTOP

"The principle of the Gothic architecture," Coleridge once said, "is infinity made imaginable." The same could be said of the modern interface. Where the flying buttresses of Chartres rendered the kingdom of heaven in stone, the information-space on the monitor embodies—"makes imaginable"—the otherwise invisible cotillion of zeros and ones whirling through our microchips. With hindsight, of course, we can see that early bitmappers like Doug Engelbart and Ivan Sutherland had it easy. They were trying to represent a modest assortment of bits and bytes, less than a floppy's worth of information. Today's interface designers are faced with a more daunting task: the gigabytes of data stored on most hard drives—not to mention the epic endlessness of the World Wide Web.

In the days before Gutenberg, cathedrals were the great signifying machines of public life. More than mere buildings, they implied a way of looking at the world, a sacred order, a sense of proportion. At a time when mass literacy was unimaginable, cathedrals served as a kind of popular text built out of stained glass and gargoyles. This sign system worked at different scales. You could read the story of Christ in the impossibly detailed stone carvings, of course, but you could also zoom out far enough to see the cathedral in relation to the town that surrounded it. That story, more than any other, was the most momentous, the most unavoidable—all the other narratives were wrapped up inside it, like subplots in a triple-decker novel. A town of shanties and thatched roofs and one-story cottages clustered around the majestic spires of the cathedral. A hundred times larger than any other built structure, and a hundred times more elaborate, the cathedral lay at the very center of the town—physically, of course, but also spiritually. You could see in a glance that this was a world anchored by religion, where all roads led back to that towering emblem of faith and submission. The organization of space— all those medieval towns wrapped devoutly around their cathedrals—not only *implied* a specific mindset, it helped create it.

This process of imagining the world through spatial organization is hardly limited to the sacred text of the Gothic cathedral. Think of the way the agora of ancient Greece—with its lively bartering and public debate—embodied the vitality and intimacy of the city-state. Or think of the consumer-society values implied in today's "edge cities": satellite communities joined by freeways and shopping malls, living spaces designed exclusively for the twin pursuits of driving and shopping. Peer into one of those shopping malls and you'll see how intimately related architecture is to the social imagination. By any reasonable standard, shopping malls break all the rules: their exteriors are bleak, unfinished, forbidding; the exits are never clearly marked; the space seems fiendishly designed to confuse. And, in fact, it *has been* designed to confuse: there's good money to be made from disorientation, as there is in the techniques of "window dressing" and "impulse buying," perfected in the first department stores of the late nineteenth century. A financially successful shopping environment is one that confuses you, that makes you lose your bearings, that keeps you walking—since more walking necessarily entails more exposure to merchandise that you might be suddenly compelled to buy. (The same principle underlies the seemingly illogical placement of escalators in multitiered malls, designed specifically to force you into making one last unnecessary loop past that Sharper Image or Banana Republic outlet.) The modern shopping mall—with its teeming, sense-saturating displays and byzantine floor plan—is the spatial equivalent of today's MTV-style advertising. In these high-velocity times, anything that disorients sells product.

In theory, these are examples of architecture and urban planning, but in practice they are bound up in broader issues: each design decision echoes and amplifies a set of values, an assumption about the larger society that frames it. All works of architecture imply a worldview, which means that all architecture is in some deeper sense political. "To imagine a language," Wittgenstein famously wrote, "is to imagine a way of life." The same is true of buildings, parks, cities—anything conjured up by the human imagination and then cast into stone. The way we choose to organize our space says an enormous amount about the society we live in—perhaps more than any other component of our cultural habits.

This, then, was the burden that confronted the first generation of interface designers working in the wake of Engelbart and Sutherland. The bitmapping revolution had introduced the concept of dataspace, but it was still mostly a tabula rasa, an empty lot waiting to be filled. What were the new information architects going to build on that real estate? A strange mix of open-endedness and limitation was contained in the question. Because the computer was by definition so malleable, capable of shape-shifting from one visual metaphor to another, it was theoretically possible for the interface to look like practically anything: a house, a factory, a movie, a diary. But the limitations of seventies technology—minuscule storage devices, sluggish microprocessors, grainy monitors—meant that flights of fancy would quickly bump up against the ceiling of hardware shortcomings. You could build anything you wanted in that new information-space—but it had to be simple, and easy to represent.

The solution that those original designers came up with still dominates the way we imagine our computers and the information buried deep within them. You can make the argument that it was the single most important design decision of the past half-century, altering not only our perception of dataspace, but also our perception of real-world environments. In an age of information, the metaphors we use to comprehend all those zeros and ones are as central, and as meaningful, as the cathedrals of the Middle Ages. The social life of that time revolved around the spires and buttresses of "infinity imagined." Our own lives now revolve around a more prosaic text: the computer desktop. Understanding the implications of that metaphor—its genius and its limitations—is the key to understanding the contemporary interface.

The story of the desktop's circuitous route to popular success—from the R&D lab to Windows 95—is by now a familiar one. Like most Silicon Valley sagas, it begins with a plucky band of outcasts and dreamers, and ends with Bill Gates conquering the planet. There are a few definitive accounts: Steven Levy's wonderful tribute to the Macintosh, *Insanely Great,* and

Howard Rheingold's visionary study of computing's early years, *Tools for Thought* (now sadly out of print). But it is worth going over the narrative again, even in an encapsulated form. After all, the history of rhetoric is at least as old as ancient Greece, but there are few metaphors in that long tradition that have changed the world with such speed and such magnitude. Computer interfaces may yet move beyond the desktop, but whatever forms eventually evolve will still owe a great debt to that original vision.

Like most technological breakthroughs, the desktop metaphor came into being accidentally, as a solution to another, unrelated problem. In its original form, the metaphor was just a throwaway analogy, a figure of speech instead of a fully realized interface. It was early 1972, and the researchers at Xerox's high-end computer science lab in Palo Alto (also known as Xerox PARC) were struggling with the legacy of Doug Engelbart's windows. The story of Xerox PARC is a strange and contradictory one. All the anecdotal evidence suggests that it was an enormously creative, intellectually challenging place, and it summoned up an immense number of high-tech innovations in less than a decade. (You could fairly say that the modern idiom of computing was born there.) But it never produced a moneymaking product in all that time—with one possible exception.

A number of Xerox PARC scientists were veterans of the Stanford Research Institute; they'd imported a selection of Engelbart's ideas about bitmapping, mice, and windows to Xerox, where they began tinkering with the original model—this time on a larger budget. One researcher in particular—a brilliant and charismatic young man named Alan Kay—was struggling with the SRI implementation of windows. It had been clear from Engelbart's first breathtaking demonstration in 1968 that windows would revolutionize the way we imagine information. But the SRI windows were clunky and two-dimensional. They didn't overlap. As Levy writes:

> While Engelbart and his Augmentation workers had pioneered the window, the partition they had in mind each staked out its own portion of the monitor. Not only was it difficult to keep straight which window one was working in, but the windows wound up competing for the extremely limited real estate on the screen. Kay's solution to this was to regard the screen as a desk, and each project, or piece of a project, as paper on the desk. It was the original desktop metaphor. As if working with real paper, the one you were working on at a given moment was on top of the pile.

Engelbart and Sutherland had endowed the digital computer with space; Kay's overlapping windows gave it *depth*. It was a subtle distinction, but a profound one. You could move in and out of the landscape on the screen, pull things toward you or push them farther away. The bitmapping revolution had given us a visual language for information, but Kay's stacks of paper suggested a more three-dimensional approach, a screen-space you could enter into. The whole idea of imagining a computer as an environment, a virtual world, comes out of this seemingly modest innovation, although it would take many years for that legacy to become visible.

As a desktop metaphor, though, Kay's overlapping windows weren't terribly convincing. Judged by contemporary standards, they weren't "visual metaphors" at all. Kay's original desktop was closer to Engelbart's mouse than to the folders and trash cans of the modern graphic interface. You didn't feel like you were sitting at a simulated desktop at all. The metaphor was just a way of explaining why some windows now seemed blocked by other windows; it wasn't an attempt to simulate a real-world desktop, just as Engelbart's mouse wasn't designed to reproduce flesh-and-blood rodents.

It was a loose analogy, but it had staying power. As the Xerox PARC team assembled its human-interface prototype over the next decade, Kay's original desktop metaphor became increasingly concrete. If the computer could take on any shape imaginable, why not have it mimic the old, analog world that it would replace. It was a kind of imaginative trade-off: if people were going to be abandoning their atom-based file cabinets and trash cans and stacks of paper, why not just relocate those things to the digital world? Part of the solution was simply functional. You could build on the user's existing strengths and aptitudes. Knowing something about how to organize a file cabinet would help you organize your digital files, just as being familiar with how trash cans work would help you delete files. The metaphors would make the user experience more intuitive, and the playful, lively graphic metaphors made the idea of using a computer much less intimidating. If you could sit at a desk and shuffle papers, you could use the machine.

And so the Xerox PARC team hammered out the first genuine desktop interface, as part of an experimental operating system called Smalltalk. Xerox never managed to do anything with Smalltalk—they bundled it with an expensive computer system called the Xerox Star that flopped spectacularly in the early eighties—but the desktop metaphor was too powerful an idea to remain trapped in a Palo Alto lab. As it turned out, the desktop was liberated by a headstrong young businessman who first laid eyes on Smalltalk during a tour of the Xerox PARC facilities. His name was Steven Jobs.

Jobs, of course, was one of the founders of Apple Computer, and he was in the market for the Next Big Thing, the technological advance that would revolutionize computing the way the original Apple II had several years before. He found what he was looking for in Smalltalk. Within two years, Apple had a desktop interface running on its Lisa machine, an expensive, underpowered product that never found a market. The following year, however, Apple released the Macintosh—the "computer for the rest of us"—with an inventive, mesmerizing desktop metaphor, one that introduced almost every modern interface element to the popular imagination: menus, icons, folders, trash cans. More than a decade later, it remains the standard by which all interfaces are judged. You can safely say that all interface enhancements since are merely variations on that original theme.

More than anything else, what made the original Mac desktop so revolutionary was its *character.* It had personality, playfulness. It displayed a masterful integration of form and function, of course, but there were also instances of *gratuitous* form, art for art's sake. Windows zoomed open. Menus flickered. You could customize the pattern of your desktop, create your own icons. The Macintosh was far easier to use than any other computer on the market, but it also had a sense of style. The awkward phrase "look-and-feel," popularized by Mac advocates, reflects just how novel this idea was. There wasn't a word to describe a computer's visual sensibility because up to that point, computers hadn't had visual sensibilities. The Mac changed all that. Staring at that undersized white screen, with its bulging trash can and its twirling windows, you could see for the first time that the interface itself had become a medium. No longer a lifeless, arcane intersection point between user and microprocessor, it was now an autonomous entity, a work of culture as much as technology.

Alan Kay had realized all of this many years before the launch of the Mac, back in those early days of overlapping windows: "The computer is a medium! I had always thought of it as a tool, perhaps a vehicle—a much weaker conception. . . . If the personal computer [was] a truly new medium then the very use of it would actually change the thought patterns of an entire generation." It was a lesson he had learned from McLuhan's *Understanding Media.* "McLuhan's claim [was] that the printing press was the dominant force that transformed the hermeneutic Middle Ages into our scientific society. . . . The press didn't do it just by making books more available, it did it by changing the thought patterns of those who learned to read." If the Mac's desktop metaphor suggested a whole new digital medium, how might the conventions and protocols of that medium change our perspective on the world? Certainly the effects would be monumental, though they would also be hard to predict.

Apple, of course, was more than ready to exploit this revolutionary language, and it launched the Macintosh with an unprecedented media blitz. The campaign itself was a milestone for a number of reasons: on a technical level, it was the first mass-media promotion that devoted as much attention to an interface as to the underlying hardware itself. As if anticipating this shift from engineering to artistry, the advertising had a pronounced—some would say strident—countercultural tone, nowhere more prominent than in the legendary "1984" ad. Routinely ranked as one of the great commercials of all time, the "1984" spot cast IBM as Orwell's despotic Big Brother—its dreary, command-line operating systems doing for most PC users what rats did for Winston Smith. In the ad—which ran only once, during the 1984 Super Bowl broadcast—the numbed masses are liberated by a torch-wielding jogger, who converts them to the Mac's user-friendly graphic interface by tossing a ball of fire at the grainy, pixelated visage of Big Brother himself. The campaign even included a suitably populist tag line: "The Computer for the Rest of Us."

But despite the totalitarian imagery of their opening salvo, the first interface wars were basically cultural in nature, more about "lifestyle choices" than anything else. PCs, with their arcane codes and hideous green-on-black monitors, belonged to the suits, to Organization Man. The Mac's playful interface spoke to a different demographic: jazzier, creative types, new thinkers and iconoclasts. Buying a Mac was an expression of individual identity, like Steve Jobs wearing T-shirts to board meetings—more of a fashion statement than a party affiliation. The computer you used revealed your personality type, not your politics. Seen from this angle, the "1984" spot looks like a mirror image of Walter Benjamin's classic analysis of fascism, an essentially aesthetic conflict coated with a thin, cosmetic layer of politics, just enough to rally the "rest of us" into action.

To a certain extent, the imaginative battle over the desktop continues to this day, though the terms of the dispute have changed. The rise of Microsoft Windows confirmed the superiority of the desktop metaphor—the primary bone of contention in the original Mac–DOS debate. With both sides now happily clicking their mice across virtual desktops, the older conflict took on new implications. Mac devotees continued to defend the intrinsic merits of their platform, but a new line of attack began to make the rounds. According to this argument, Apple's products were worth supporting simply because they were the only things keeping Microsoft from total domination of the operating systems market. In the digital world, where the need for compatibility propels consumers toward a single industry standard, this sort of power rippled out quickly into other fields. Microsoft's stranglehold on operating systems translated into massive advantages in other software markets: business applications, or home entertainment. Apple—the story went—was the one thing standing between Bill Gates and a full-throttle monopoly. Rooting for the Mac was no longer just a lifestyle issue; it was a way of standing up for fair competition and free markets. The stakes were political, not cultural—a battle against consolidated power rather than an act of personal expression.

Already bound up in political and cultural dispute, the desktop soon entered into legal imbroglios as well. In 1995 the Justice Department began an investigation into Microsoft's newly announced online service, the Microsoft Network (MSN), for alleged antitrust violations. The main thrust of the complaint—filed by several of the company's competitors in the online service business—revolved around the placement of a single icon on the Windows 95 desktop. That icon served as a gateway to the MSN registration screens, enabling users to subscribe without ever leaving the desktop level of the Windows 95 environment. At the time, industry analysts predicted that this high-profile icon would allow MSN to suck up

surd to us now—like asking, "Are the typewriter keys
r do you also need a ribbon?" The same article goes
te an expert who bemoans the simplified idiom of
 icon: "When you combine the mouse and the icons
 new language, a language in which there are no
 you want to open a file you position your hand on
nd you move it and the arrow moves to the file. You
 and that's the action. So all the verbs take place by
 That's very appropriate for gross positioning or
asks, but as soon as you get into anything that's
 than that you've simply got to revert back to a
terface." There's some subtlety to the observa-
 still revolves around this bizarre opposition
ns and the menus, as though the two were
ists rather than natural allies.

gtime advocates of the graphic interface will
 wry vindication in these excerpts, as well
mories of battles with DOS snobs in the late
 willful myopia and unwarranted prejudice
hole host of disastrous appraisals of new
 passages also make you think of count-
reviews of pathbreaking works of art,
 assimilated into the canon of great art.
, or Stravinsky's *Rite of Spring*.) In both
ental inability to see the thing itself in
-whether we're talking about a novel or
tics saw him as a pornographer, not a
s that rioted during the first perfor-
certainly didn't think they were being
hing comparable happened with the
early eighties. You can see in that

nine million subscribers in its first year of operation. (Its
biggest competitor, the four-year-old America Online, had just
over three million.) Figures like this led the other online services to charge foul play, contending that Microsoft was
exploiting the strengths of its operating-system market share
to squelch competition in the budding online industry.

No matter which side you agreed with on the
issue, it was worth pointing out that the entire skirmish—
involving several massive corporations, numerous pundits both
online and off, and the Justice Department of the United States
of America—came down to a tiny little icon, winking inno-
cently at us from the cathode-ray glow of the monitor. An icon!
It was hard in the midst of all that sturm und drang to
remember the debates we had ten years ago about the Mac's
icons. The central question, at that point, had been: were they
too cutesy? Did they make the Mac seem too much like a toy? If
the desktop itself entered into the discussion, it usually had to
do with the merits of the checkerboard pattern versus the standard gray. (Again, aesthetics over politics.) Fast-forward a
decade and suddenly the question of whether an icon was on or
off the desktop was a matter of national import, worthy of discussion by Janet Reno and the *New York Times* op-ed page. Put
in the simplest terms, what we had was a significant investigation by the highest law enforcement agency in the land, triggered almost entirely by computer interface design. If there
was ever an argument for the power and the scope of the
desktop metaphor, it was to be found in this investigation. Bury
that pesky little icon three levels deep in the file directories
and—voilà!—the subpoenas disappear. Plant it squarely on the
desktop and the Feds start busting down the door.

•　　•　　•

Looking back now, with more than a decade's worth of hindsight, what strikes you about the early days of the desktop metaphor is how many people resisted the idea, and how many simply didn't get it at all. The viability of the graphic interface is so far beyond question now that it's difficult to remember that there was ever a dispute about it. But if you sift through the original reviews of the Mac and the Lisa—as well as of the pseudo graphic interfaces that appeared in other software packages about that time—you can't help but be struck by how hard a time the critics had wrapping their minds around the new paradigm.

Some of the reviews of the graphic interface struck the ridiculous real-men-don't-do-windows chord that reverberated through corporate America in the nineteen-eighties, as in this wag from *Creative Computing* magazine: "Icons and a mouse will not make a non-literate person literate. Pointing at pictures can last only so long. Sooner or later you must stop pointing and selecting, and begin to think and type." The opposition now seems completely out of place to us, accustomed as we are to the way spatial metaphors can *augment* thought—but to those first critics, the visual language seemed like child's play, or a cartoon. Other reviews missed the point altogether, dismissing the Mac as a tool that only artists and designers would have use for, as though the machine's major innovation was MacPaint's spray can and not the interface itself. Consider this editorial from *Forbes,* dated February 13, 1984:

> [The Macintosh's] best features are for computer
> novices: MacPaint, a program that creates graphic
> designs of stunning complexity, and MacWrite,
> a word-processing program that goes to ingenious

lengths to set up
writer. Both ar
"mouse," whic'
user's touchir
not aimed at
manager ha
of MacPai
time wri
about de

The
brilliance of tho
breathtaking, of
graphic interf
There's not ev
organizing i
doesn't mak
use would
a puzzlin
interfac
for gr
promi

of
ar
t

seems a
enough,
on to qu
mouse an
you have
verbs. . . . I
the mouse a
click on that
hand action.
very simple t
more complex
menu-driven i
tion here, but
between the ic
somehow antago
probably find som
as a few darker me
eighties. The mix o
brings to mind a w
technology. But thes
less early, scornful
works that were late
(Think of *Ulysses* here
cases there is a fundar
the proper framework—
an interface. Joyce's cr
novelist, and the crowd
mance of *Rite of Spring*
subjected to music. Some
desktop metaphor in the

Forbes editorial the same sensibility that caused the stampede out of Stravinsky's debut. A new sound descends upon you for the first time, and all you hear is noise. The music part escapes you altogether. Likewise, a new way of representing information comes across your path, but all you see is "simplicity," a graphics program, or a child's toy.

There's something satisfying in reading through these old dispatches, but we should also be a little humbled by this historical record. The first reviews of the Mac and its brethren testify to the conceptual limitations that come out of working under one paradigm and then struggling to adapt to another. There are invariably blind spots and omissions in these transition points, things that seem intuitive with hindsight but were almost impossible to grasp at the time. More than a decade after the Mac's launch, with the triumph of Microsoft Windows having confirmed the basic merits of the graphic interface, it's tempting to think of our current understanding of the interface as a purely enlightened one, free of prejudice. Alas, the desktop metaphor has as many limitations and conceptual blind spots as its command-line predecessors. Only these restrictions come from being too faithful to the original metaphor itself, extending the original desktop into more fully realized 3-D spaces, into office buildings and living rooms. The conceptual failings of the mid-eighties resulted from an inability—or an unwillingness—to see the power of the desktop metaphor. The failings of the present day come from taking that metaphor too literally.

Le Corbusier once described a house as "a machine for living in." He may have got it the wrong way round. Our machines now seem on the verge of becoming houses, populated by perky animated characters who guide us through the clutter and remind us of daily chores, like a fifties sitcom housewife sculpted out of binary code. This, at least, is the vision of Microsoft's Bill Gates, whose 1995 software package Bob promised a living room on every desktop, with cartoonish assistants personalized for each member of the family. Developed under the supervision of Melinda French (now Mrs. Gates) and tellingly code-named Utopia, Bob emerged in the spring of 1995 with a stunning amount of media fanfare, particularly given the mediocre sales that followed its launch.

Like many of Microsoft's interface "advances," Bob borrowed heavily from preexisting innovations—in this case, General Magic's 1994 operating system for handheld computers, Magic Cap. Both Bob and Magic Cap projected the user into a familiar, three-dimensional environment styled after a real-world equivalent. Bob employed the extended metaphor of a living room that could be redecorated according to the user's sensibility. Magic Cap worked with an office metaphor, complete with hallways, reference rooms, and a virtual "downtown" that the user visits when connecting to online services like CompuServe or America Online. New programs appeared as objects in the room; install a spreadsheet on your hard drive, and an adorably obsolete mechanical calculator appears on your shelf space. In the world according to Chairman Gates, programmers seemed fated to become interior decorators, scattering bric-a-brac and binary potpourri across computer monitors worldwide.

The paradoxical thing about these hypermetaphors was that they weren't metaphorical enough. In the *Poetics,* Aristotle defined metaphor as the act of "giving the thing a name that belongs to something else." The crucial element in this formula is the *difference* that exists between "the thing" and the "something else." What makes a metaphor powerful is the gap between the two poles of the equation. Metaphors create relationships between things that are not directly equivalent. Metaphors based on complete identity are not metaphors at all. In traditional interface design, a computer "window" bears a kind of superficial resemblance to a real-world window, but it's the differences between the two that make the metaphor a successful one. (We'll look into this further in the next chapter.) We obviously can't layer our kitchen windows atop one another, nor can we scroll through the vista they provide. There's a necessary distance between the real and virtual window that makes the analogy useful to us.

Bob and Magic Cap were in the business of eliminating that distance. As Alan Kay writes:

My main complaint [about modern interfaces] is that *metaphor* is a poor metaphor for what needs to be done. At PARC we coined the phrase *user illusion* to describe what we were about when designing user interfaces. There are clear connotations to the stage, theatrics, and magic—all of which give much stronger hints as to the direction to be followed.... Should we transfer the paper metaphor so perfectly that the screen is as hard as paper to erase and change? Clearly not.

The interface environments of Bob and Magic Cap weren't so much metaphors as they were simulations. In the original Macintosh design, your computer desktop functioned *like* a real-world desktop, just as your file directories functioned *like* real-world folders. Bob did away with those comparatives: your computer *is* a living room, and your life in front of the monitor should be lived according to your habits and tastes away from computer, down to the decor of the space itself, which you can transform from dusty Victorian to postmodern sleek with a single mouse-click.

Somewhere along the way the good faith of user-friendly metaphors had been replaced by the hysteria of total simulation. The reasonable desire for analogies between the digital and the organic had given way to an all-encompassing quest for a pure fusion of the two. We already have enough living rooms and hallways to go around; we don't need them replicating across our monitors as well. A hallway is a perfect example of the limitations of real space, limitations that the computer needn't trouble itself with. In the off-line world, you can't build rooms, Borges-style, with an infinite number of connections to other rooms, whereas a computer hooked up to the Internet can handle links to millions of other spaces effortlessly. There's something perverse in this total deference to user-friendly simulation, like building a word processor that faithfully reproduces a mechanical typewriter, complete with stuck keys and worn-out ribbons. It's user-friendly, all right, but who wants that kind of friend?

Microsoft explicitly targeted Bob at digital neophytes and technophobes, the sort who stand on chairs screaming at the sight of a computer mouse. Power users were never part of Bob's imagined demographics (though Magic Cap was targeted at more computer-savvy mobile professionals). But even if Bob were to remain an entry-level interface, its celebrated accessibility has some real drawbacks, drawbacks that have everything to do with the spatial metaphor employed by the software. The original desktop metaphor was just loose enough to avoid feeling restrictive or excessively bureaucratic. You weren't fooled into thinking that you were working within a fully realized virtual office, which is one reason that Apple was able to market the Mac as a liberation from dull corporate conformity. (If the desktop metaphor had been closer to a simulation, the campaign would have seemed absurd.) Bob's living-room metaphor, on the other hand, feels unnervingly *safe,* like a bland, lifeless, gated community where perfectly manicured hedges line empty streets. Bob represents the domestication of the personal computer, in the pejorative sense of the word, turning the miraculous shape-shifting capacities of these machines into a dulled repetition of everyday, household reality. The real magic of graphic computers derives from the fact that they're *not* tied to the old, analog world of objects. They can mimic much of that world, of course, but they're also capable of adopting new identities and performing new tasks that have no real-world equivalent whatsoever. People who get hooked on computers get hooked for this reason. They don't become high-tech junkies because their machines remind them of their Rolodexes; they're junkies because their machines do things they never thought possible. Interface design should reflect this newness, this range of possibility.

And computer novices, more than anyone, need to understand that potential. Entry-level interfaces should explain why a computer is unlike anything that's come before it, which is precisely what makes it so fascinating and so powerful. Interfaces such as Bob give us the comforting illusion of domestic life, the same old living rooms recast for the digital age, with a few animated puppets to liven things up. But it's a sedative and not a jump start, the mother's-little-helper of interface design, calculated to block out everything that's so promising and unpredictable about the medium. Interfaces styled after Bob or Magic Cap may yet produce millions of domesticated computer users, dutifully transcribing their recipes into digital cookbooks, but how many of those users will venture beyond those clean, well-lit spaces into the more complex and revolutionary world where computers aren't answerable to the familiar dictates of interior design? Bob isn't so much a gateway to the information age as it is a landing pad, a way of cushioning the blow for novices edging their way over the high-tech precipice. And when they find that their new life on the other side of the monitor looks exactly like their old one, will they really be inclined to stay?

The other problem with the domesticity of Microsoft's Bob is that the imagined space is a profoundly anti-social one. It conceptualizes the infosphere as a private home, sequestered from the outside world. The only contact with other "people" comes in the form of those ridiculous cartoon characters, those agents and info-butlers. There's a strange sense of agoraphobia hovering over this world, as if the happy-go-lucky, Disneyfied interior was just a roundabout way of blocking out the shocks and the turmoil of public life. This might have been reasonable in the old days of stand-alone desktop computers, but in the age of the Internet, using an interface that doesn't offer some vision of public life can seem less like a cutting-edge exploration through information-space and more like a visit to Miss Havisham's.

This gets to the heart of the desktop metaphor and its broader implications. Organized space implies not just a personal value system—as in the religious order of the Gothic cathedrals—but also a type of community. This is true of architecture and urban planning, and it is also true of interface design. The cramped and crooked side streets of Paris up until the late nineteenth century (still visible in parts of the Latin Quarter and the Marais) invoked a human scale of neighborhoods and face-to-face contact, more like village life than that of a great metropolis. (The crowded conditions also created public health problems, of course, as in the 1832 cholera epidemic.) The city had an improvised, organic quality to it: streets wrapped haphazardly around each other, neighborhoods evolved unpredictably. There were a few regal exceptions to this rule, buildings or public environs laid out by princes or priests, but for the most part the city was a great celebration of self-organization, a design etched out by millions of small-scale, local decisions, with no master planner in sight.

This principle of self-organization implied a very specific understanding of what urbanism was about. The city was seen as a system driven from the bottom up, created by the countless daily acts of individuals following their routines: bartering, chatting, building, tinkering. If those crooked Parisian streets were the embodiment of that self-organized mentality, the novel was its reflection. (Flaubert and Balzac leaned relentlessly on the plot device of one character accidentally bumping into another character on the street—it has the same kind of canonical stature that the Main Street shoot-out does in spaghetti Westerns and John Ford flicks.) But that particular spatial model of the city had its antagonists. When the first great urban developer, Baron Haussmann, razed those older neighborhoods to build the grand boulevards of modern Paris, his construction crews destroyed more than just buildings. They also destroyed a long tradition of imagining how cities worked. Where the older Parisians saw the streets as a way of facilitating the random, casual encounters of public life, Haussmann conceived of the streets in purely functional terms: first, as a way of moving people from one place to another as efficiently as possible, and second, as an epic, built-world tribute to his patron, Napoleon III. The broad, straight lines of the Rue de Rivoli and the Boulevard Saint-Michel were alms laid on the altar of modern efficiency (not to mention the Empire itself). In the twentieth century, Haussmann's vision of urban life has become the conventional wisdom. You can see it in the edge cities of Los Angeles and Phoenix, in Eisenhower's Interstate Highway Act of 1954, in Robert Moses's notorious Cross-Bronx Expressway. As Le Corbusier—who yearned to repackage the Marais as an immaculate grid of buildings—once suggested, the modern street is a "machine for creating traffic." If a city happens to get in the way of that machine, too bad for the city.

There is a lesson here for interface designers, as they look for ways to expand the desktop metaphor into the public life of the Internet. We used to hear a great deal about how computers were creating a generation of asocial geeks, more comfortable with their gadgets than with real people, but the rise of bulletin-board communities like the Well and ECHO—not to mention the Web itself—has changed all that. In the past few years, a more encouraging trend has become apparent to most people who have spent time online. Instead of being a medium for shut-ins and introverts, the digital computer turns out to be the first major technology of the twentieth century that brings strangers closer together, rather than pushing them farther apart. Most of the major innovations of the past hundred years have made it progressively easier to avoid contact—and particularly conversation—with people who aren't colleagues, or family, or friends. The automobile created the isolated cloisters of suburbia; the telephone and the television kept us firmly implanted in our domestic spaces; even the public life at the cinema unfolds under a vow of silence. The last major technological revolution that brought strangers closer together was the cotton gin and its industrial descendants, which relocated millions of workers from the sparsely developed countryside of Europe and the eastern United States and crammed them into the tenements and assembly lines of factory towns like Manchester and Lowell. The Internet is once again allowing strangers to interact with one another, though this time without the violence and the drudgery of the Industrial Revolution.

There's something deeply encouraging in this rediscovered public life, but much of it is still speculative. Much of it, in fact, will depend on the interfaces dreamed up in the next few years, interfaces designed to represent communities of people rather than private workspaces. Interestingly enough, the early returns suggest that these interfaces may well use 3-D environments—not unlike the living rooms and offices of Bob and Magic Cap—to their advantage. The question is whether these new environments will end up looking like the gated communities of Los Angeles or the more open-ended, improvisational street theater of traditional urban life.

On the screen you are a yellow sphere, sporting Groucho glasses and a bow tie, floating in a dark, austere hallway. The space is littered with a dozen other yellow spheres, similarly accoutered. At first glance, the scene suggests a costume ball at a tennis clinic, at least until you notice that the spheres are talking to one another, in cartoon-style text balloons that pop up beside them at regular intervals. Every minute or so one of the spheres departs for another room, and every minute or so a new one arrives to join the conversation. The banter, to be sure, is more than a few notches below Algonquin Round Table levels, with an undue emphasis on the phrase "where you from?" and various garbled obscenities. But the experience is hypnotic nonetheless.

This surreal tableau belongs to Mark Jeffries's extraordinary 1995 software creation, The Palace. Each yellow sphere represents a computer user logged into the Internet; a given Palace room might contain users from Malaysia, Prague, and Peoria, chatting each other up in real-time conversation. Global chat, of course, has been around for several years—in the much-ridiculed chat rooms of America Online, or the typing jam sessions of Internet Relay Chat (IRC). But the Palace introduces a critical spatial element to what had previously been an exclusively textual affair. Traditional chat rooms were rooms only in the loosest sense—more like scripts really, with each new line scrolling down a text-only screen. The Palace projects those conversations into an ambient space; the original Palace contained banquet halls and boudoirs, stairwells and drawing rooms. Users can roam freely through this architecture; some tend to settle into a favorite space, while others enjoy hopping from environment to environment. In the chat rooms of AOL and IRC, users concoct a persona for themselves primarily through their screen names, giving them limited opportunities for self-definition. The Palace software lets you craft a visual presence that other users experience when they encounter you in a room: you can don a range of gay apparel—from the Groucho glasses to a glittering tiara—or you can dispense with the yellow sphere altogether and fashion yourself Dick Nixon or Pamela Lee Anderson.

Like so much of today's interface culture, The Palace's most innovative feature has remarkably low-tech roots: in the popular "sporting room," you can challenge another user to a game of chess, sliding rooks and pawns across a virtual board while other users loitering around the room look on. This strikes me as being one of the small miracles of contemporary interface design—not for any radical innovation, but for the way it simulates the casual sociability of pickup chess matches in Washington Square Park or the Tuileries: a few strangers tossed together by chance and shared interests, exchanging a few words over a game of chess, while others heckle and second-guess around them. Up to now, social interfaces have been relentlessly textual, nothing but a parade of words on a screen; The Palace interface adds a whole new dimension to the virtual community: the more visual, improvised theater of town squares and urban parks, pickup softball games and watercooler banter.

Software like The Palace makes me think that the whole "Web surfing" metaphor may prove inadequate for the social meanderings of most netizens. Real-world surfing, after all, is an exceedingly solitary activity; in its traditional usage, the Web surfer is seen battling the ceaseless waves of information flow, without much regard for the other surfers out there navigating the same channels. (We will return to the limitations of the surfing idiom later.) The Palace, on the other hand, suggests a much more pedestrian metaphor: Baudelaire's flaneur, the "man of the crowd" drawn to the tumult of the nineteenth-century boulevard, drawn to the "kaleidoscope of consciousness" found among the teeming masses prowling those metropolitan streets. The chance encounters of The Palace interface are a sign of things to come: social interfaces that approximate the thrill, and the unpredictability, of casual encounters in a more textured space, shaped by the physical presence of those around you and the possibility of interaction that goes beyond a few polite epithets. On your first visit to The Palace, you can't help sensing that something powerful is working its way into existence through those floating orbs, the sort of rich, vibrant interaction that strangers once enjoyed on the streets of our cities and that they may yet enjoy again on the virtual boulevards of cyberspace.

And yet the promise fades quickly. On most of my visits to The Palace, I find myself thinking happily of Baudelaire and Washington Square Park until I start paying attention to what's being said. Despite The Palace's emphasis on the *social* possibilities of interface design, the conversations taking place within those oak-paneled walls leave a great deal to be desired. Consider this representative exchange:

Guest 872:	et tu viens souvent en france
Prince Thiago:	^wait
Dollar:	clean
:Steven:	ah oui
Guest 688:	non jamais
Guest 702:	HELLO homofobiazns how is it sceezin
Guest 872:	pourquoi
:Bob:	hi 541
Guest 880:	flexing your chest
Dollar:	where is you're palace
:Bob:	sure are you?
rock:	Salut dollar
K-MAN:	what are you doing owner

Guest 872:	)kiss
Dollar:	hello
Guest 880:	to painful
K-MAN:	what are you doing owner
K-MAN:	what are you doing owner
Guest 688:	pas assez d'argent
K-MAN:	what are you doing owner
:Bob:	where are you 541?

I've heard a few online denizens make the case for this language as a kind of digital-age free verse, a verbal stew of disconnected phrases and libidinal outbursts, something Burroughs might have scissored together on a slow night in Lawrence. But it reminds me of graffiti, and graffiti of the worst kind: isolated declarations of selfhood, failed conversations, slogans, tag lines. You don't really see a community in these exchanges; you see a group of individuals all talking past one another, and talking in an abbreviated, almost unintelligible code. Most real-time chat is like this, of course, as a quick visit to AOL's conversation spaces will make clear, but somehow the flatness of the language, its evasiveness, seems more pronounced when projected onto the vast quarters of The Palace. At least on AOL the bare-bones visuals—a line of text scrolling down the screen—feel somehow commensurate with the general pitch of the conversation. Wandering through the banquet halls and grand staircases of The Palace, you can sense the staccato dialogue being dwarfed by the surroundings. (It's a little like renting out the London Symphony to play a few bars of "Happy Birthday.") There's something wrong in the *scale* of the experience, and you can't help but wonder if the problem will only grow more exaggerated as our virtual environments become more lavish, and more true to life.

Certainly, the most successful online communities to date have been text-only affairs, almost without exception. The Well and ECHO, Parent Soup and The Book Report, HotWired's Threads, Howard Rheingold's Electric Minds: almost every thriving digital gathering place has anchored itself squarely in text. (Most of these communities still have the visual sensibility of DOS circa 1989.) For all the furnishings, The Palace doesn't feel like a *place* yet. It feels more like some massive back-lot set with a dozen tourists and passersby improvising lines awkwardly under the floodlights. You don't feel at home in this environment; you feel lonely—or worse, you feel the loneliness of being trapped in a room with a handful of loners, reaching out over the wire for those abrupt, meaningless entreaties: "what are you doing owner? what are you doing owner? where are you 541?" It's not the act of seeking companionship and camaraderie over the Net that disconcerts here. The citizens of ECHO and the Well regularly establish rich and lasting relationships over the modem, relationships that eventually meander their way into face-to-face contact. Mediated conversations are not by definition shallow—think of the telephone, where "reaching out to touch someone" can take the form of a heart-to-heart with a loved one or a sex-line chat with the recorded voice of a porn star.

The medium itself may be capable of creating both lifelong friendships and flimsier, pickup-room encounters, but clearly the spatial metaphor (or lack thereof) has an enormous influence on the type of community created. And this gets to the most perplexing part of the matter: almost without exception, the leading examples of digital sociability didn't require a spatial metaphor to make their communities happen. For the most part, the social fabric of cyberspace is still stitched together by the gossamer thread of text. There are plenty of Silicon Valley players betting that this will change, as 3-D software grows more commonplace and users grow comfortable navigating through more realistic environments. But perhaps the text-driven model will have a longer shelf life than the soothsayers think. It is conceivable that, by the end of the next decade, we will arrive at a consensus that larger virtual communities—communities made up of hundreds of involved citizens—may simply exceed the representational capacity of *any* spatial metaphor. If the depth of shared experience is the yardstick by which you ultimately measure your community— and it's probably as good an index as any—then I must admit that I have a hard time imagining a better platform for community building than the traditional, text-based bulletin-board system utilized by ECHO and the Well (along with many Web sites). These acclaimed "electronic salons" are information-spaces in only the loosest sense of the term, and yet they don't appear to suffer at all from the lack of an environmental metaphor. (Electric Minds actually built a Palace as a customized extension of their text-driven Web site, but it is nearly always empty.) In my experience at least, there is more shared wisdom in a single thread on the Well than there is in a hundred Palace gatherings or 3-D chats.

But perhaps this is yet another case of the blindness that accompanies every significant reworking of the interface medium. If *Forbes* could see only the child's-toy sensibility of the Mac desktop, then surely we shouldn't be so quick to write off the inane, abbreviated banter of The Palace. I have no doubt that people will develop new forms of conversation better suited to these environments, and that surprisingly powerful interactions will emerge out of them. (Think of the way the edgy formality of early phone conversations metamorphosed into the more probing, engaged marathon sessions of your average adolescent.) Even today, the early experiments with onscreen "avatars"—digital representatives that stand in for you in the virtual space, like the glowing orbs of The Palace— suggest that future interfaces may allow for more physical, more gestural expressiveness. No doubt that line of research will eventually yield a convincing, intuitive interface for a gathering of four or five people, a design that relies heavily on a spatial metaphor to bring those individuals closer to one another. But I am not so confident about communities of hundreds of people, or thousands. There the spatial metaphors begin to break down, and the asynchronous, text-only posts of most bulletin-board systems seem more appropriate, more enabling.

Of course, not every online gathering place is judged by the standards of conversational depth. As it turns out, the one domain that has successfully extended the original desktop metaphor into three dimensions is that of video games—most notably the blood-and-guts, "first-person shooter" genre of Doom, Marathon, and Quake. Interface design probably comes closest to architecture in these programs, as game players scurry through their chambers of gore, toting machine guns and blasting at everything in their path. Given the core audience of teenage boys, there's no denying that the carnage is an important part of the appeal, but what catches the eye initially about these games is the visual rush, the vertigo of moving through a textured onscreen space at high velocity. (The rise of motion sickness as a regular side effect of playing these games should be a sign that something significant is in

•

i
n
t
e
r
f
a
c
e

c
u
l
t
u
r
e

the works here.) The pleasure of these games is as much the pleasure of mastering a space, learning to navigate though it, as it is the pleasure of shooting things.

But the architectural element in these games goes well beyond merely occupying a space. The shareware libraries on the Net teem with new levels for Doom and Marathon, digital buildings constructed by end-users and decked out with the requisite futurist-medieval paraphernalia. Among gaming aficionados, trading custom-designed environments is a ritual as commonplace as the proverbial neighbor borrowing the proverbial cup of sugar. Sure, the virtual spaces being swapped back and forth won't give the Centre Pompidou a run for its money—most of them look like B-movie versions of *Alien* or *Excalibur*—but the very nature of the exchange is telling in its own right. After all, these aren't baseball cards or GI Joe dolls we're talking about; these are little worlds, little environments. Dreaming them up in the first place is a legitimate form of self-expression (even if it is constrained by the generic bloodlust of Quake and Doom), and the idea of sharing these worlds with other gamers suggests a whole new model for community building, where the exchange between individuals no longer simply takes place *within* a space. Instead, the space serves as content, not context. The bartered game levels function like sentences in this peculiar conversation, the back-and-forth of varied worldviews jostling for supremacy or approbation. You want your Quake level to impress your fellow gamers, or seduce them—the way the decor of your apartment or your office is designed to make an impression on any visitor who happens to stumble into it. The architecture of that virtual space doesn't *frame* the conversation—it's a central component of it. We're used to communicating with our friends and family by sending them snapshots or sketches or tape mixes, but in the future we will reach out to those around us by sharing virtual environments. Constructing elaborate palaces as tokens of affection used to be the exclusive province of royalty and multimillionaires. Perhaps now the gift of built space will become more commonplace as a gesture of friendship or affection, once the software moves beyond the high-octane fury of Doom and Quake.

The wonderful thing about these trading rituals is that once you've shared a Quake level, you can always gather together a group of friends or strangers and go hang out in it. Id Software—the company behind both Doom and Quake—designed the latter for the Internet from the ground up, enabling dozens of players, each logged onto the Net from a different locale, to battle one another in Quakespace. Watching one of these multiple-player sessions doesn't exactly reinforce the conventional wisdom of the Net bringing strangers closer together, since what these strangers are doing is slaughtering one another, and the only dialogue takes the form of Schwarzenegger-style taunts. ("Hasta la vista, baby!") But there is no question that the architectural metaphor has made the gathering possible, even if its core activity doesn't live up to the communal hype. That alone suggests that spatial metaphors of the original desktop will expand into more vividly realized environments over the next few years, environments designed specifically to accommodate gatherings of individuals separated geographically. The real question is whether these environments will be good for anything other than simulated carnage. Already, there is talk about co-opting Quakespace for more peaceful activities. Id Software kept specifications for designing levels as open-ended as possible, with the hope that end-users or other software companies might design virtual worlds that were not exclusively fixated on bloodshed. There's no reason why a Quake level couldn't be designed to accommodate a game of hide-and-seek—or, for that matter, a weekly Emily Dickinson reading group. It sounds improbable, of course, but stranger things have happened in technohistory.

If more enlightened communities do eventually sprout in Quakespace, there will be a certain measure of irony to this transformation. After all, the first spatial metaphors to find their way into the computer interface were mistaken for video games, and it took years for the DOS snobs and command-line devotees to accept the computer desktop as anything other than a child's toy. Perhaps with the rise of the "first-person shooter," the sequence will be reversed: a seemingly mindless video game will end up transforming our sense of information-space, the way the desktop metaphor did twenty years before. The original advocates for the graphic interface spent countless hours keeping their creations separate from the superficial world of video games. Tomorrow's desktop metaphors—and particularly those metaphors designed to represent online communities—may very well prove to be émigrés from that world. What was once seen as a threat to the desktop metaphor may well turn out to be the most fertile breeding ground for its successor.

three

WINDOWS

Thanks to Microsoft's lavish advertising budgets, the window is now shorthand for the wide array of innovations that make up the modern interface. Forget about the mouse pointer, and the desktop metaphor, and the menu bar—the history of interface now neatly divides into two epochs: pre-windows and post-Windows. But what exactly does a window do? This seems like a simple enough question, given the you-can't-live-without-it hype that billowed out of the Windows 95 launch. The simplest, and most tautological, answer—the one you'd expect to hear from the marketing departments—is that windows make our computers easier to use. But why should a window-driven interface be easier to use than a text-driven one?

You'd think the answer would have something to do with our innate capacity for visual memory. After all, this was the great selling point of the original graphic interface. If Matteo Ricci could recall an entire biblical treatise by spatializing the language, turning words into architecture, then surely the transformation of bits and bytes into a virtual space must augment our data-recollection skills. The relationship seems simple enough: spatial information is easier to navigate than textual information, and windows are just a tool for seeing that space, like a looking glass or a microscope.

Although the explanation sounds plausible, it doesn't correspond to the way most of us use windows in our everyday computing lives. The window actually has little to do with remembering where something is, the way we might remember where we last saw the car keys, or the route to a friend's house. Spatial mnemonics are an essential part of the modern graphic interface, of course, but they're mainly concentrated in things like the menu bar and the location of desktop icons such as the trash can. Successful interfaces—the original Mac design, later versions of Windows—insist that these elements remain unchanged from application to application, for precisely this reason. Every Mac user knows how to cut and paste because he or she knows *where* the copy and paste commands are—in the upper-left-hand corner of the screen, under the "Edit" menu item. The knowledge becomes second nature to most users because it has a strong spatial component to it, like the arrangement of letters on a QWERTY keyboard. As with the original typewriter design, the consistency of the layout is as important as the layout itself. (In the case of the typewriter, of course, consistency is everything, since the keys are actually arranged to slow down the typist.) Spatial memory works only if the objects you're trying to keep track of remain anchored in one place. There's no use memorizing where the trash can is if it keeps meandering around your desktop.

Windows are more fluid, more portable. You can drag them across your screen, resize them with a single mouse click. They're designed to be malleable, open-ended. Most computer users are constantly tinkering with their windows, making them bigger or smaller, pushing them off to the peripheries of the desktop or bringing them into focus. What good is our visual memory when it's dealing with a device that moves around so much? No good at all, as it turns out. Like many of our file-management tools, windows rarely work in the service of spatial memory, despite what the interface evangelists will tell you. Think about the way a modern, windows-driven desktop interface deals with storing your documents. The official line is that you remember where you put a given file because you're thinking in terms of "where" in the first place. (In a command-line system, the question would be, What sequence of letters do I type to call up this document?) In other words, the graphic interface endows the file with spatial coordinates, giving it the spatial properties of a file on a real-world desktop.

But this conventional wisdom is somewhat misleading, precisely because the window itself is so flexible. Even the most devoted windows advocate still thinks of his or her files in textual terms better suited to command-line interfaces such as DOS and Unix. To understand this, you need only pay attention to your thought processes when you're fumbling around with your file-management software, looking for a stray document. In a purely spatial system, you would think to yourself: I feel like the file was over on the left-hand side of the screen, down a few layers. But in reality, what you think is this: I'm pretty sure I put it in the "Things to Do" folder, but maybe it's in "Unfinished Business." In other words, you're organizing information *textually*, in terms of categories that you've defined yourself. The spatial dimension is just an illusion, or the illusion of an illusion. We pretend to ourselves that we're remembering "where" we put the file, but what we're really remembering is the name of the folder that contains it.

This is a major distinction, and one that is too easily glossed over in discussions of the modern interface. Sometimes your "Things to Do" folder will appear as an array of icons at the very top of your screen; sometimes it will appear as a smaller window down at the bottom. It all depends on your passing whims—and the arrangement of other windows on your screen. That adaptability is part of the window's appeal, of course, but it makes it difficult indeed for our much-touted spatial memory to do its magic. (The one exception to this is the case of files that have been placed directly on the desktop, bypassing the window altogether. These icons can develop genuinely spatial attributes, making them easier to find—though most people prefer not to have a desktop cluttered with too many icons.)

The easiest way to grasp the textual limitations of a windows-based file system is to spend some time with a genuinely spatial alternative. For several years, Apple has been tinkering with a 3-D file-management interface originally code-named Project X. (Its latest incarnation is called HotSauce.) Whereas most interfaces represent documents as files nested within folders, all planted squarely on the computer's virtual desktop, HotSauce imagines your data as a kind of galaxy, with documents and folders floating like so many planets against a dark backdrop. At first the user confronts a vista of, say, six or seven terrestrial bodies looming in the night sky, each representing a file directory. The scene is not altogether unlike the top-level view in a traditional graphic interface. Only in this environment, clicking on one of the items doesn't pop open another window; it zooms in the entire frame closer to the data-planet you've selected, as though you were approaching it in some kind of spacecraft. As you move closer to this object, a ring of other files and directories slowly comes into view, like an array of satellites orbiting the main planet. Translated into the language of the desktop interface, these satellites turn out to be the files and folders contained within the higher-level folder. You can click directly on the objects that represent files to open the corresponding document. Objects that represent folders reveal their own satellites as you inch closer to them. The user genuinely navigates through the dataspace, zooming in and out, veering left and right in search of the right planet.

It may be a long time before we replace our mundane window-driven environments with HotSauce's NASA-style interface, but Apple's prototype serves as a useful reminder of how small a role spatial memory plays in the modern interface. As a case study of sorts, I used HotSauce as a replacement for my file system for a few days, just to see what the experience would be like. The initial explorations were enormously amusing—more like playing a video game than organizing my files—but the thrill quickly subsided into irritation as the navigational limitations became apparent. It took too much energy and attention to move through the space; I ended up thinking more about how to steer the device than about the data I was looking for. By the end, I was happy to return to the more prosaic, two-dimensional world of my desktop, with its text-driven file system. But a day or two with HotSauce was enough to catch a glimpse of what a genuinely spatial system might feel like. At a few, enthralling moments, I found myself groping around for a familiar document and thinking: It's back there somewhere, up and to the left a little, about two or three planets deep. For a second or two I was thinking in purely spatial terms, zooming in and out of my own private dataspace. For those few moments, there was a hint of liberation in the air, the promise of things to come.

• • •

If windows don't harness the potential of our spatial memories, then what are they good for? In the early days of the graphic interface, as the computer world struggled to make sense of this new computing revolution, the usual explanations revolved around being able to see two documents at once. You could do a quick compare and contrast between two drafts of a text document, or finesse a few numbers in one window while a pie chart adjusted itself accordingly in another. That was a tantalizing prospect in its day, although a handful of command-line applications already offered "split screen" features. For most users, though, the real benefit of the window comes not from being able to see two documents at the same time but from being able to switch back and forth between those documents with a single click of the mouse. The window turns out to be a way of visualizing what programmers call a mode switch.

     In an average day working at a computer, chances are you switch back and forth between dozens of different modes without thinking twice about it. Imagine a mode as being a rough appraisal of what your computer is doing at that very moment. (The word *mode* has a precise technical definition, but for our purposes we'll use the term in a broader sense.) You have a mode for creating a new text document; a mode for editing an existing spreadsheet; a mode for rearranging a file directory; and a mode for changing your system preferences. In the old days of the command line, you had to initiate each of those modes by typing in an obscure sequence of letters, and the line between each mode was clearly drawn. If you typed one sequence you would enter into the directory-altering mode; if you typed another you'd be allowed to tinker with your system preferences. This required prodigious feats of memorization, of course, and it was easy to lose track of which mode you were in. The whole system was fiendishly counterintuitive: it was as though each time you wanted to scratch out a note to someone on a piece of paper, you had to punch in a key combination to unlock your pen.

What Doug Engelbart and the folks at Xerox PARC realized was that you could take these arcane, command-driven modes and replace them with windows. The spatial properties of the window weren't a mnemonic device, a way of remembering where you'd put things. They were a way of representing *modes*—and, more important, a way of switching back and forth between modes. The top-level window would represent the active mode ("edit the contents of this directory"), while beside it a nonactive window tempted you with another mode ("change your system preferences"), while beneath both those windows another lurked, this one offering to edit a word-processor document. You could shuttle back and forth between modes effortlessly, by clicking on the appropriate window. The illusion was so successful, in fact, that the whole idea of "modes" has dropped out of mainstream computer parlance to be replaced by "windows."

That shift from modes to windows was a massive advance in ease of use—so massive, in fact, that it is now difficult to imagine a digital world without windows. Creative transformations of this magnitude tend to have secondary effects on those of us living under their spell, particularly when the conventions are so familiar, so second nature that they become transparent to us. (Think about the way those "memory palaces" shaped the structure of Dante's *Inferno*.) For cyber-philosophers like Sherry Turkle, the windowed imagination is emblematic of our larger "postmodern" condition:

the unified field of traditional post-Enlightenment thinking fractured out into a hundred different points of view, each of them equally valid. The passage from the fixed system of the command line to the more anarchic possibilities of the window follows the same route traveled by Western philosophy: from the stable, unified truth of Kant and Descartes to the relativism and ambiguity of Nietzsche and Deleuze. The window, for Turkle, is a way of thinking in multiplicities, as all good post-modernists are supposed to do. "Multiple viewpoints," Turkle writes, "call forth a new moral discourse.... The culture of simulation may help us achieve a vision of multiple but inte-grated identity whose flexibility, resilience, and capacity for joy comes from having access to our many selves."

It's easy to imagine the potential objections to this line of logic. For high-tech skeptics like Sven Birkerts, the window is less a matter of "multiple selves" and more a matter of attention deficit disorder, yet another sign of the culture's resistance to the slow, contemplative pace of the traditional novel. An aesthete like Birkerts, trained to celebrate poetic ambiguity, wouldn't object to the basic philosophical premise of Turkle's argument; it's the digital *means* to that conceptual end that he would have a problem with. Surely we don't need so many modes, buzzing simultaneously in every corner of our computer screens, to appreciate the insights of relativist philos-ophy. (Nietzsche, after all, did pretty well with the linear, print-on-paper technologies of his day.) More conservative cultural critics like the late Allan Bloom and Dinesh D'Souza would simply reject the suspiciously Continental idea of multiple truths or perspectivism. They might agree with Turkle's assess-ment of the way virtual windows shape the contemporary mind. The difference is that Bloom and D'Souza would consider that shaping deplorable, yet another baleful influence—like television or rap lyrics—seducing the youth of America.

What all these various positions agree on, though, is the underlying premise that windows lead inexorably to a more fragmented, disconnected experience of the world. And I'm not totally convinced that this a reasonable a priori assumption. There is no doubt that the transparent mode switches of a windows-driven interface allow us to multitask more easily with our computers, though most of the time our routines involve discrete sequences, where we concentrate separately on one task after another rather than managing them all at once. Still, the onscreen space *has* grown more layered and multiplicitous, and of all the interface innovations, the window has played the most important role in making this possible. Turkle is right to point out that more is happening on our screens, but of course "more" is a relative term—which is why any account of the computer's newfound complexity must also acknowledge how dull and one-dimensional it once was. "Multitasking" is a digital-age term, but the basic process of concentrating on several things at once is hardly a Silicon Valley invention. People have been multitasking for centuries—if not millennia—as anyone who has cared for an infant or done homework in front of the TV will testify. If anything, the digital computer kept us abnormally focused on single tasks during the command-line years. The rise of the window simply restored us to our usual fragmented state: the sort of multitasking we labor through every morning, reading the paper while getting dressed, all the while keeping an eye on the bacon and eggs on the stove. The window metaphor was a genuine liberation for most users, but what it liberated were innate skills that had long been suppressed by the awkward mode switches of the command-line regime. The window didn't create a new consciousness—it just let us apply our existing consciousness to the information-space on the screen.

Not all the issues raised by the modern windows-driven interface have to do with postmodernistic sociology or our fragmented sense of self. Some of the most intriguing—and unexpected—side effects of the windows revolution turn out to be legal and ethical disputes, revolving around journalistic principles and intellectual property rights. But to understand those relationships, we need first to take a look at how the window has matured since those heady days at Xerox PARC and SRI.

Of all the basic building-block metaphors in the rhetoric of interface, the window has evolved the least over the past twenty years. Not since Alan Kay stumbled across a way of layering his windows has the metaphor seen a significant innovation. Consider the competition: the desktop metaphor has ripened into 3-D offices and town squares; the mouse pointer has spawned onscreen avatars; the menu bar is now supplemented by floating palettes and context-sensitive wizards. Our windows shrink and snap into place with more fluidity than they did in the olden days of Smalltalk, but they remain the same windows. The flourishes may have changed slightly—grayscale scroll bars, minimize buttons—but the basic mechanism has stayed the same.

But if the window hasn't changed much in a decade, the view certainly has. The scroll bars once flanked spreadsheets, text memos, black-and-white drawings. Now they sit idly beside DVD video streams and virtual town squares. You'd think the increased complexity of the landscape would justify new windows, new lenses through which to view the infospace. But the innovations have been sparse and underpublicized; almost none of them have reached commercial success, though there are some promising candidates on the horizon. One such candidate comes from Xerox PARC, the wellspring for much of the modern interface. Its inventors call it the "magic lens."

The magic lens works by inverting the modus operandi of the traditional window. As is true for most interface conventions, the inner workings of the window are fiendishly difficult to describe, despite being immediately intuitive when you see them in operation. (This, of course, is part of the point of the graphic interface—you have to *see* it to believe.) But it's worth walking through the basic laws of this digital universe, if only to better understand how those laws might be improved upon. For the most part the traditional window remains stationary, gazing out onto a dataspace that scrolls up or down, or from side to side. The contents move, not the form. You can drag the entire window across the desktop, of course, but the contents are lugged along with it when you do.

The first time you encounter a scrolling window, there's a sense of depth to it. The window appears to look out onto a dataspace that continues beyond the borders of the window itself. You imagine your document extending down below the screen, scrolling like a TelePrompTer, or like one of those moving panoramas from the nineteenth century. But the illusion quickly wears off. The window starts to feel more two-dimensional, more like a piece of paper than a portal. The view-space appears to flatten out, to the point where the window and the data contained within the window merge. You don't feel that you're looking *through* a window at something. You feel that you're looking at the window itself.

The point here, of course, is that the luster of the original metaphor has worn off, become literalized. This process is as old as language itself. The world of words is littered with these dead metaphors, poetry that has long since ossified into prose. Do we think of sunlight when we describe a book as "illuminating"? Do we imagine fingers and thumbs each time we contemplate the hands of an analog clock? Of course not. The familiarity of the terms has erased their rhetorical value, like a coin whose face is rubbed out from wear and tear, transforming it from currency to simple metal. (The analogy is from Derrida's wonderful essay, "White Mythologies.") The window has suffered the same fate. We no longer think of our virtual windows as *analogs* of the real-world version. They're a species unto themselves.

The magic lens is an attempt to restore some of the optical metaphor to the modern graphic interface. The lens resembles nothing so much as a magnifying glass—it hovers above your documents, a metawindow of sorts. You look through it at the document below, though what you see is radically transformed by the lens itself. You could easily build a magic lens that worked the way a traditional magnifying glass works. Drag it over a complex illustration and within its frame you see an enlarged version of the image. Outside the lens, the image remains at its normal size. This alone would be a useful tool for some image-processing programs, where zooming into the pixel-space can prove disorienting. The lens allows you to peer into the detail work without losing a sense of the whole.

But a magic lens is more than just a magnifying glass—that's what makes it magic, after all. Whereas a magnifying glass only enlarges, a magic lens can do countless tricks, some visual, some textual, some oral. It's up to the interface designer to decide which attributes of the main document the lens should transform. Instead of size, the lens could punch up the info-resolution of a document. Imagine a road map with the bold arteries of major highways spanning the image. Drag the lens over certain regions, and it reveals the intricate lattice-work of minor roads leading away from those highways. A full-scale map at that resolution would be much more difficult to decode; the broad strokes, the major relationships, would be lost, and you'd be left with a meaningless jumble of back alleys and culs-de-sac.

Like the original window, the lens sounds like a tool for revealing things, for opening up new data landscapes, but in practice it turns out to be more useful for shutting things out. The lens is a tool for discriminating. It filters, and in doing so it keeps many things opaque. The lens acknowledges that surplus information can be just as damaging as information scarcity. The highway map example illustrates this principle nicely. You don't need a map that delineates every road in the state. It'd be useless, far too rich in information, like the famous 1:1 cartography in the Borges story, the map of the county that *is* the county. What you need is a map that will let you block out all this excess information and concentrate on the regions that require more depth. You can create a similar effect in some programs by zooming farther into the information-space, the way you do with Photoshop's magnifying glass, but that zoom invariably makes it harder to see the big picture. You have to shuttle back and forth between magnifications to take in all the data. The lens lets you do both at the same time.

The magic lens is one of those puzzling new ideas that *sounds* profound when you first hear about it but never quite lives up to your expectations. The metaphor is so convincing, and so intuitive, that you feel you've stumbled onto something that will soon integrate itself into your everyday interface routine, like the first time you double-clicked on an icon and a new window spun out to fill the screen. But the prototypes and case studies of the magic lens at work can ring a little hollow. They're electrifying to see, of course—the lens drifting across a document like a Ouija board—but the real-world applications can seem a little, well, redundant. The brilliant CD-ROM adaptation of Leonardo's Leicester Codex (produced by Bill Gates's multimedia lab, Corbis) probably counts as the most celebrated use of the magic lens. In the CD, Leonardo's famed backward script can be mirrored by one magic lens and translated into English by another. You pull the lens across the yellowed, brittle pages—or their digital reproductions, to be more precise—and the words beneath the lens snap into recognition. When I first saw Corbis demonstrate the product at a conference in 1996, the translation lens got a standing ovation. You could hear the gasps in the crowd as the lens delivered up the English rendering. But after the applause died down, you had to wonder how different it would have been had the lens been replaced by a single button that translated the entire screen with one click. It might not have been as nifty to look at, but would the information have arrived with any less efficiency?

We'd do well to remember, though, that the history of computing abounds with interface enhancements that sprouted new uses once they left the R&D lab. Because the magic lens can reveal such a wide array of data types—typeface specifications on a desktop publishing document, traffic congestion on an urban planner's map, contaminated regions on a nuclear power plant's blueprint—it's entirely possible that the lens will eventually become as second nature to us as ordinary windows are today. But like so many contemporary interface filters, its value will come less from what it reveals than from what it keeps hidden—that great, teeming potentiality of infinite data lurking within most digital texts. In *Labyrinths,* Borges wrote, "I cannot walk through the suburbs in the solitude of the night without thinking that the night pleases us because it suppresses idle details, just as our memory does." Anyone who has spent any time meandering through the suburbs of the infosphere will tell you that we too are awash in idle details, and could use a little suppression. For that very reason, the magic lens may yet find a significant role in interfaces to come.

If the magic lens's mainstream success is still open to question, Engelbart and Kay's ancestral windows do have one clear descendant in the contemporary interface family. But it is more like a gene splice or a phantom limb than a biological heir. Instead of summoning up a whole new species of window, the modern interface has taken to drawing and quartering the original beast, dividing it into subwindows, or frames.

A frame is something like the picture-within-a-picture feature that you'll find on the televisions of sports fanatics and aspiring media critics. Any computer user who has tinkered with a word processor that allows you to divide your document into parts will recognize the device immediately. It's a window, in short, that offers more than one view—like the split-screen polyphony of the prom scene in Brian De Palma's *Carrie.* Conventionally, modern interfaces respond to a request for another perspective by calling up another window, another view out onto the dataspace, separate but equal. Frames deal with these requests by carving up the existing windows into discrete units. A single window might contain several frames, each contemplating a different region of the infosphere. In the most rudimentary examples—the split-screen word processor, for instance—one frame might display the beginning of a document while the other frame focuses on the closing paragraph. But in the more complicated environments of the Web, frames take on more challenging—and perplexing—representational tasks. Their use has been at the center of a number of heated debates, some of which have spiraled out into lawsuits—yet another example of how seemingly innocuous interface tools can trigger political and legal firestorms.

The frame is an example of something like Darwinian "exaptation" at work in the world of technology. In biology, exaptations are variations on the usual variations of natural selection. Unlike *adaptations,* which are changes in the organism that respond to environmental conditions (the giraffe that evolves a longer neck to reach the leaves of acacia trees), *exaptations* are novel, unexpected applications of these new traits. Evolution selects for longer necks in giraffes so that they can strain farther upward in search of food, but sometimes the eventual applications are not part of the original selection process. The wing, for instance, evolved independently as an extension of the reptile's webbed forelimbs. Evolution, in other words, was dutifully selecting for stronger, more aquatic limbs, and somewhere along the way, it stumbled across the capacity for flight. As Darwin puts it in his *Origin of Species:* "An organ initially constructed for one purpose . . . may be converted into one for a wholly different purpose."

If the fossil record abounds with such unlikely conversions, the high-tech record is positively swarming with them. Design a tool to solve one problem, and soon enough you'll find another problem that the tool can solve. More often than not, it's a problem you had barely noticed before, because there were no tools on the horizon that could deal with it. The digital spreadsheet was originally designed to record financial information, like the rigid double-entry books of paper-based accounting, but it quickly became a tool for *modeling* the numbers, for experimenting with the possibilities, building simulations. Jaron Lanier's virtual-reality goggles began as a quest for a visual programming language but soon morphed into an entertainment device. The Web itself is a kind of large-scale exaptation. Originally designed as a local filing system for academic research, it became a mass medium almost overnight—conveying news, soap operas, diary entries, soft-core porn, and almost anything else you can imagine to an audience of info-consumers worldwide.

The frame has followed a comparable evolutionary path. It began as what programmers call a "kludge," a short-term fix to a more fundamental problem. In the early days of the Web, the technical language used to describe digital pages—HTML—had a notoriously limited vocabulary. There was a direct, one-to-one correspondence between each page and the window that represented it to the user; if you scrolled down the page, the elements at the top would disappear off the top of the window. You had to scroll back to see the original information. This was perfectly intuitive, of course, particularly during the Web's infancy, when most of the information came in the form of text documents. But as sites grew increasingly complex over that first year, Web designers sought out new ways to map the complexity, enabling users to better navigate through sites. They designed button bars and site maps, similar to the "you are here" signs in shopping malls, and grafted them onto each page. Like the menu bars of the graphic interface, these design elements were implemented with consistency in mind—the infospace might be disorienting, but you'd always have a toolbar of options at the bottom of the screen in case you got lost. The only trouble was, the navigational devices kept disappearing from view every time a user scrolled through the window. You couldn't anchor them into place the way the menu bar was anchored to the top of the screen in the Mac interface.

Financial pressures also influenced the creation of frames. In the fall of 1994, *Wired* magazine's online creation—HotWired—was launched, with a full suite of sponsors on board promoting their wares through slender "banner" advertisements perched at the top of every page. The real estate left something to be desired: agencies accustomed to concocting full-page spectacles had to pack their brands into a space smaller than a playing card. But the potential rewards of an online campaign were tempting: users could click on the banners and transport themselves to the sponsors' own Web sites, where they could browse through merchandise or fill out marketing surveys. Once again, scrolling windows were the enemy. Sites like HotWired were selling eyeballs, but those eyeballs kept drifting down the page, pushing the banner advertising off the screen. You can't click on something you can't see.

In mid-1995, frames were introduced into the HTML parlance to deal with these problems. Instead of creating an entirely new specification for toolbars or permanent ads, the HTML consortium—and the programmers at Netscape who refined the standard somewhat—went with a simple, open-ended solution. You could divide the window into separate frames, each pointing to a different document. A frame at the bottom of the screen might point to a site map, while the rest of the window displayed scrolling text. In actuality, the map and the document would be separate pages, with separate addresses (or URLs, in Web jargon). But the illusion would be of a single page with a navigational image lashed permanently to its base. The same went for advertisements—the user could roam aimlessly through the site, but the banners remained securely moored at the top of the window. Technically, you were looking at two distinct pages, but the real-world effect was of viewing a single page that happened to have an advertisement plugging away above the main text.

So far so good. The frame evolves as an *adaptation* to environmental conditions—mainly the need for architectural clarity and advertising dollars. But where does the exaptation come in? As it turns out, the HTML implementation opened up an intriguing new possibility for site designers. If a window could now point to two different pages at the same time—each one showcased in a separate frame—it could also point to two different *sites*. Nothing in the code dictated that both pages come from the same source. Instead of linking to another site, you could call it up directly within a frame. You could build a Frankenstein document, stitched awkwardly together from secondhand parts: a frame pointing to the White House page planted next to a McDonald's promotional site, all flanked by a banner advertisement stolen from Time Warner's Pathfinder Web presence.

But was it really stolen? That remains the great question posed by Web frames, and it is one that will only grow more treacherous over the next few years. It gets to the heart of the intellectual property issues raised by digital technology— though once again, it emerged through a simple interface enhancement, a minor fix. Think about it this way: if I take the contents of *Newsweek*'s Web page and put them up on my Web site, I am clearly violating the magazine's rights. But if I simply put a link from my site to *Newsweek*'s site I'm completely within the law—in fact, I'm doing *Newsweek* a favor (a modest one, perhaps, but a favor nonetheless). A frame lies somewhere in the hazy middle ground between these two extremes. Let's say I create a page on my Web site divided into two frames: one points to *Newsweek* and one points to entries from my journal that I have lovingly reproduced online. Above both frames, in clear, bold digital type, is the title: Steven Johnson's writings. Now, am I illegally reproducing someone else's intellectual property, or am I simply referring to it? Keep in mind, my document merely *points* to the *Newsweek* URL. The actual text and code reside on *Newsweek*'s servers; I haven't directly copied any of the material. But to a user casually browsing through my pages, the difference will be hard to detect. All he or she will see is a page with two frames, one of which appears to be my ramblings; the other appears to be a cover story from a major periodical. Is this the equivalent of my recommending a good article to a passing stranger? Or is it the equivalent of color-Xeroxing a handful of *Newsweek*s and selling them on the subway?

The short answer is, we don't know. We don't know because intellectual property law, as it stands now, doesn't have a language to describe the new realities of digital information—and particularly digital information piped through the infinite relays of the World Wide Web. On the most basic level, the problem here is that our intellectual property laws don't know how to deal with *windows,* with metaforms that hover in that strange zone between medium and message.

If I show you a copy of *Newsweek* through my personal window, is that like selling a tape of the World Series without the "express written consent of Major League Baseball"? Or is it like inviting friends over to watch a ball game from an apartment that happens to overlook Comiskey Park? On a technical level, the difference between linking to another site and reproducing it within a frame comes down to a few stray words of HTML code, but there's an entire lifetime of intellectual property-law disputes contained within those words.

The issue came to a head in February 1997, when a group of prominent media companies—including the *Washington Post* and *USA Today*—filed suit against a small Web start-up called Total News. The plaintiffs contended that Total News had violated their copyrights by unlawfully reproducing articles that had originally run on their Web sites. For people outside the technology world, the case must have seemed strange. Total News didn't deny reproducing the articles, but it strenuously denied the accusation of copyright infringement. How could this be?

The answer, of course, revolved around frames. Total News had hit upon an intriguing business model, the sort conceivable only in a digital world. What it offered was a kind of electronic newsstand or clipping service. The presentation was straightforward enough—a window divided into three frames. A frame flanking the left-hand side of the window listed the names of ten major news providers, including the *Post* and *USA Today*. If you clicked on the *Post*, the newspaper's front page washed slowly into the main frame in the window. During all this, as you moved from site to site, a frame wrapped along the bottom of the window steadily broadcasted advertisements. It was a classic case of media parasitism, a metaform that created value by pointing to other, "real" content. Total News created an audience by gathering together all the major national papers in one site—or, more precisely, gathering together *pointers* to all the major national papers. Once it had attracted a large enough audience, it could make money by selling ads. Since the cost of stitching together those pointers was almost nothing—you could build the same page yourself with the simplest HTML tools—the revenues would grow steadily along with the audience, while the expenses remained negligible.

It was an intriguing business model, but was it against the law? It will take a long time to find out, given the nebulous state of digital intellectual property law. But whatever the outcome, it's worth pointing out that the entire affair—the pilfered newspaper content, the ad banners, the prepackaged list of major sites—would have been unthinkable without frames. A small interface embellishment created to make Web sites more "user-friendly" had launched a multi-million-dollar lawsuit involving some of the most powerful media conglomerates in the world. The Total News case may be one of the first such cases triggered by virtual windows, but it's not likely to be the last. At least in this example, the ethical questions posed by the computer interface are a matter of public debate. We have not always been so lucky. In fact, some of the most troubling collisions of journalistic ethics and information design have gone virtually unnoticed, partly because the technology is so new and partly because it is in the nature of windows to be transparent.

Consider, for the sake of argument, the following hypothetical scenario: It's early September of an off-season election year, and the *Washington Post* is ramping up its coverage of a tight Virginia Senate race between an incumbent with long-standing ties to the D.C. establishment and a feisty upstart, understaffed but full of fervor. The *Post* has kept the story at a steady burn for the past two months, with the occasional flare-ups appearing above the fold, and the occasional op-ed leaning toward one of the two candidates. There are other races to cover, of course, but the story looks to be a regular page-one item for the months to come—and probably the lead come election day.

And then, suddenly, the *Post* announces a special offer to its readers: if they promise to vote for the incumbent in the Senate race, they'll have the paper delivered to their doorsteps free of charge for the rest of the year. That's right— free. *Libre*. Gratis. *Service compris*. The Virginia challenger cries foul, puts a few constitutional lawyers on the payroll—and then immediately starts lining up deals with other local papers.

Sound appalling? It should. Implausible? Not so fast. A comparable lapse in journalistic ethics occurred in the fall of 1996. But in this case, public indifference to the offense was more thunderous, and more dispiriting, than the crime itself. The real culprits were not the bloated, woozy doyens of the Beltway high life but rather the sleek starlets of the digital age: the West Coast info-magnates and their chroniclers in the business press. And it all revolved around a seemingly harmless computer window.

Here's the real story. In August of 1996, after a long and frustrating battle against Netscape, Microsoft announced an ambitious new plan to link its Web browser to certain "content providers," following the obligatory late-nineties corporate dictum: if you can't beat them, synergize them to death. Meanwhile, the Dow Jones Web masters had decided that it was time for folks to start coughing up for the *Wall Street Journal*'s interactive edition. The libertarian credo—"information wants to be free"—might play well under Montana's big sky country, but try telling it to the bond traders.

Hence the synergy. The *Journal* was looking for a way to erect barricades without completely alienating an online population accustomed to free Web sites. Microsoft wanted a way to differentiate its Web browser from the competition, a "value-add" that had more to do with the information relayed by the browser than the browser's actual features. It's a move every MBA knows by heart. You can see the same logic at work in the fast-food chains every time your burger arrives in a luminous box gilded with Disney's latest canonical cartoon. (Even the "nourishment providers" have to serve up a little content every now and then.)

And so on August 13, Microsoft and the *Journal* announced an agreement whereby users of Microsoft's Internet Explorer browser would be welcomed with open arms at the *Journal*'s Web site, while the rest of the online community lined up at the box office to shell out the $29 annual fee. The main portal into the *Journal*'s online presence soon sported a decorative "Download Internet Explorer" icon, complete with the Microsoft logo. On an otherwise austere, text-driven page, the icon looked like Super Mario pasted into a Ken Burns documentary. Aesthetic decisions like this may seem esoteric, but remember: the *Journal* still refuses to use *photographs* on its print pages.

At first glance, the arrangement doesn't seem equivalent to tampering with a federal election. In another industry, the analogy might be overstated. But consider the larger context of the story, and the players involved. The

Journal is the paper of record for corporate America, just as the *Post* commands the political coverage in the D.C. area, if not the entire nation. And in the world according to the *Wall Street Journal* circa 1996, there was no story more interesting and more consequential than that of Microsoft's quest for the Internet. It was, undeniably, a great story, far more arresting than your everyday Virginia Senate race, a kind of Faust for the pocket-protector set: the tottering giant of the information age, undercut by the very technological forces it had unleashed on the world, struggling to regain dominance over its industry. And in the many rivulets of strategy and circumstance that contribute to the broad stream of that story, there was no current stronger than the Web browser wars.

    A browser, of course, is a Web-friendly derivation of the original window, a way of seeing the unseeable. Install a browser on your hard drive and the teeming alternative universe of the Web comes into view, like mitochondria springing to life beneath a microscope's gaze. The Web's data is designed to be viewed by many different types of browsers (what the computer world—always quick with the religious metaphor—calls being "platform agnostic"), but, alas, some browsers are more equal than others. For most of 1996, Netscape's Navigator product had dominated the Web-browsing market, a dominance that had already transformed itself into several billion dollars, after Netscape pulled off the most successful initial public offering in Wall Street's history during November of 1995. But Microsoft had been pushing its Internet Explorer product fiercely in recent months, and threatened to incorporate the browser into the basic environment of the Windows 97 operating system.

    It was an incendiary mix, particularly given the statistical torpor of the *Journal*'s usual beat: the geek-chic

CEOs—Gates and Ballmer in one corner, Andreessen and Barksdale in the other—duke it out while the fate of the digital age hangs in the balance. And unlike some other Big Picture stories in the *Journal*'s purview, this one happened to reduce down to a single number. A percentage, in fact: the market share of the Explorer browser, jittering away over the months like a seismograph—or, for that matter, like the popular vote. And in the browser wars, as in elections and earthquakes, a five-point swing is a major story.

This is where the ethical questions go beyond the usual hand-wringing over corporate synergy and product tie-ins. Put simply, the *Journal* claimed to provide an objective account of events in the high-technology world, and yet it offered to waive the subscription fees of readers who willingly altered those events by selecting Microsoft's product over Netscape's. Given the number of *Journal* readers with Web connections, and the popularity of the *Journal*'s Web site, there's no question that the partnership boosted the Explorer browser's market share at the expense of Netscape. The *Journal* says, in effect: we're a reputable source for coverage of the race between Microsoft and Netscape, but if you'll just cast your vote for Microsoft, you needn't bother to pay for your subscription. It was a clear violation of journalistic ethics, and yet the whole affair passed by almost unnoticed. Why?

There were several reasons. The old guard of the print establishment remains, for the most part, baffled by the arcane new idiom of cyberia. Bits, bytes, banners, browsers—who can keep it all straight? And given the steady flow of press releases and wire stories, adding up to a veritable corporate Kamasutra of cross-promotional entanglements, it's almost impossible to separate the unprincipled bedfellows from the merely strange. On the West Coast, of course, the techno-libertarian weltanschauung made it hard to object to any corporate partnership—even one involving Bill Gates. (Besides, in this case the real ethical breach belongs to the *Journal,* not Microsoft.)

But there was also the novelty of the browser itself. Over the past few centuries, the newspaper medium has gradually established a few exceptions to its general principle of "objectivity," exceptions that are by now intelligible to a wide audience: op-ed pages, for instance, and advertising. A Web browser is fundamentally different from these older conventions, since it exists on a higher level than the raw information itself. The browser is a metaform, a mediator, a filter. It is a window looking out into dataspace, separating user and information, but also shaping that information in all sorts of subtle and not-so-subtle ways. It has no real equivalent in the print world, since there the reader consumes the information directly, newsprint to optic nerve, with no mediation at all, save the usual early morning grogginess or caffeine high. A digital version of the same paper is by definition more malleable. The current crop of browsers primarily differentiate themselves by supporting various multimedia standards that alter how the data looks online, but a number of new products are on the horizon that actually filter through information based on *semantic* variables. (We will see more of these in the "Agents" chapter.) Today's browsers alter the look-and-feel of the data they convey; tomorrow's will alter the meaning of that data, by emphasizing certain stories over others, or by punching up sections that are particularly relevant to the reader.

Whatever this new creature becomes, it should be clear that it does not belong to the older categories of advertising copy and op-ed pages. If there is a lesson to be drawn from the *Journal*'s ill-conceived alliance with Microsoft, it is that the ethical standards of print journalism do not always translate well into digital code, and in some cases a whole new set of principles—governing the relationship between content and filter, data and mediator—may be required. The *Journal*'s special offer for Internet Explorer users ended in January 1997, but unless we initiate a serious public discussion of these issues, carried on in a language that the "digitally challenged" can understand, it will be only the beginning of a larger, and more complex, crisis for journalists everywhere. Publishers like to talk about the "separation of church and state," the cordon sanitaire that divides advertising and editorial. Perhaps with the rise of these new information filters—these windows and browsers and lenses strung out somewhere halfway between medium and message—journalism will have finally found its own third estate.

The windowed imagination doesn't simply pose a threat to existing journalistic conventions. It also presents an opportunity for a new kind of journalism, one closer to the metaforms and information filters we saw in chapter 1. Reporters and op-ed writers and television pundits have always offered a window on the world, a distinct point of view on the day's events. The windows of information-space serve up a strangely literalized rendition of that long-standing tradition, only what the new journalism offers is a view that looks out at other views, a window that opens onto another window. In today's journalistic universe, we gravitate naturally toward news gatherers and op-ed writers whose opinions interest us, or whose reporting strikes us as being the most accurate and informative. Sometimes that reputation for excellence centers itself in a single "paper of record," but most of the time, we pick and choose, depending on the field being covered. We read the *Journal* for business coverage, the *Times* for foreign affairs, the *Post* for Washington politics, and the *New Yorker* for cultural reviews. These subtle discriminations function as a kind of information filter, one that we build for ourselves every time we reach for one information source instead of another.

Over the next decade, this stitching together of different news and opinion sources will slowly become a type of journalism in its own right, a new form of reporting that synthesizes and digests the great mass of information disseminated online every day. (Clipping services have occupied a comparable niche for years, though their use is largely limited to corporate executives and other journalists.) Total News gives us a glimpse of what these new information filters will look like, but the site neglects the defining element of a successful metaform, which is an actual editorial or evaluative sensibility. Total News simply repackages the major online news services indiscriminately; it may be a more convenient format, but it adds nothing to the actual content of the information. More advanced news "browsers" will include a genuine critical temperament, a perspective on the world, an editorial sensibility that governs the decisions about which stories to repackage. The possibilities are endless: a filter for left-leaning economic and political stories; a filter for sports coverage that emphasizes the psychological dimension of professional athletics; a filter that focuses exclusively on independent film news and commentary. The beautiful thing about this new metajournalism is that it doesn't require a massive distribution channel or extravagant licensing fees. A single user with a Web connection and only the most rudimentary HTML skills can upload his or her overview of the day's news. If the editorial sensibility is sharp enough, this kind of metajournalism could easily find enough of an audience to be commercially sustainable, given the limited overhead required to run such a service.

This should be cause for encouragement for anyone interested in a more populist model of journalism, and it is evidence once again of the wide-ranging influence of the digital window and its offshoots (the browser and the frame). Of course, these news-filtering applications were almost impossible to imagine back when Alan Kay was hammering out the layered look-and-feel in the mid-seventies—primarily because the whole idea of online publishing was at least a decade away. The window made it possible to think of journalism as a filtering process, a view that looks out onto other views, but you couldn't do anything with that vision until the major news organizations had found their way onto the great interconnectedness of the World Wide Web. The window made it possible to see information-space in a new light, but the hyperlink let us stitch that world together into a more coherent shape. It is to the link that we now turn our attention.

four

LINKS

If the mid-nineties battle over Ebonics taught us anything, it's that the lexicon of popular idiom and slang is never quite what it appears to be on the surface. Colloquial speech gets a bad rap, but more often than not slang is where language happens. The influx of new terms and intonations keeps the word-world lively. (Think of the way Yiddish has enlivened urban American conversation.) But slang doesn't necessarily rely on phonetic innovation. Sometimes the most influential buzz-words come into popularity as crossover hits, appropriations—the way NASA jargon infiltrated the national vocabulary after the moon landing. Popular slang has borrowed heavily from the digital idiom in recent years: the ubiquitous "cyber-" prefix, the broad assault of "spamming." (I've heard more than a few friends punctuate an especially profound statement with the exclamation, "Click on *that!*") It's only fitting that Silicon Valley should serve up these new turns of phrase; having borrowed a handful of metaphors from the analog provinces, the digital idiom is now returning the favor.

But not all slang translations do justice to their new environments. Like the desktop metaphors of the graphic interface, colloquial phrases that hop from one context to another run the risk of confusing matters. The familiarity of the phrase has an initial value, the way the desktop helped millions of users acclimate to the idea of information-space. But the analogy invariably has its limits. There are always threshold points and variations that separate the metaphor from the thing itself. Sometimes the gap is so wide that the translation obscures more than it reveals—like a desktop metaphor so convincing that we neglect the computer's miraculous aptitude for shape-shifting. In both interface design and popular slang, some migrations from one context to another just aren't worth the trip.

So it is with the verb *to surf* and all its variations: Web surfer, cybersurf, surfing the digital waves, silicon surfer. Not only are the iterations inane, but the concept of "surfing" does a terrible injustice to what it means to navigate around the Web. In this case, it's not the allusion to literal surfing that leads us astray—though the laid-back, Jeff Spicoli 'tude of most real surfers hardly corresponds to the caffeinated twitch of your average Webhead. What makes the idea of cybersurf so infuriating is the implicit connection drawn to television. Web surfing, after all, is a derivation of channel surfing—the term thrust upon the world by the rise of remote controls and cable panoply in the mid-eighties. Those aimless excursions across the landscape of contemporary TV—roaming from infomercial to C-SPAN to news bulletin to cartoon—were so unlike anything that had come before that a new term had to be invented to describe them. Applied to the boob tube, of course, the term was not altogether inappropriate. Surfing at least implied that channel-hopping was more dynamic, more involved, than the old routine of passive consumption. Just as a real-world surfer's enjoyment depended on the waves delivered up by the ocean, the channel surfer was at the mercy of the programmers and network executives. The analogy took off because it worked well in the one-to-many system of cable TV, where your navigational options were limited to the available channels.

But when the term crossed over to the bustling new world of the Web, it lost a great deal of precision. Web surfing naturally came to be seen as an extension of the television variety, the old routine of channel surfing dressed up in high-tech drag. With that one link of association, a whole batch of corollary attributes wrapped themselves around the hapless Web surfer. We knew from countless pop-psychological treatises and op-ed pieces that channel surfers suffered from many ailments: they were prone to attention deficit disorder and ill-inclined to perceive causal relationships; they valued images over text, but rarely watched anything for more than a few minutes at a time. These were the pathologies of the channel surfer, and they were dutifully transferred to the channel surfer's Web-based kindred as soon as the phrase was coined. Thereafter, the two activities—roaming through the mediasphere via remote control and following links through cyberspace—became variations on the same theme. Neo-Luddites like Sven Birkerts and Kirkpatrick Sale offered up lamentations on the new generation of surf-addled zombies, bewitched by the disassociative powers of the remote control and hypertext, oblivious to the ordered, moral universe of linear narrative. Gen X advocates like Doug Rushkoff built up successful consulting careers by championing the improvisational skills of today's media-savvy "screenagers."

But both the Luddites and the GenXers were seriously misguided. Web surfing and channel surfing are genuinely different pursuits; to imagine them as equivalents is to ignore the defining characteristics of each medium. Or at least that's what happens in theory. In practice, the Web takes on the greater burden. The television imagery casts the online surfer in the random, anesthetic shadow of TV programming, roaming from site to site like a CD player set on shuffle play. But what makes the online world so revolutionary is the fact that there *are* connections between each stop on a Web itinerant's journey. The links that join those various destinations are links of association, not randomness. A channel surfer hops back and forth between different channels because she's bored. A Web surfer clicks on a link because she's interested. That alone suggests a world of difference between the two senses of "surfing"—a difference that contemporary media critics would do well to acknowledge.

Unfortunately, the media critics are only half the problem. Silicon Valley itself has proved to be just as inept when it comes to the new explorations of hypertext, most egregiously in recent start-ups like Netscape and Excite that owe their billions to the Web's overnight success. That success is a direct measure of the power and the promise of hypertext—all those links of association scattered across the infosphere—and yet most Web-specific start-ups have studiously ignored hypertext, focusing instead on the more television-like bells and whistles of grainy video feeds and twirling animations. There is no little irony in this state of affairs: companies that rose to prominence on the shoulders of hypertext ignore the links as soon as they go public, as though hypertext were just an afterthought, a passing fancy. You can see this strange neglect as yet another case of Silicon Valley striving for the Next Big Thing, its dialectical quest for ever more enthralling technologies. But you can also see it as a case of sawing off the branch you're sitting on.

This indifference to hypertext stems in part from the ill-suited adaptation of the "surf" idiom. The allusion to

TV flattened out the more engaged, nuanced sensation of pursuing links, made it harder to see the real significance of the experience, which then made it harder to imagine ways in which it could be improved. That neglect is no small matter. Consider just this one statistic: near the middle of 1996, Netscape and Microsoft released new versions of their respective Web browsers, setting some sort of informal record for the most rapid-fire software upgrades in history. These new versions between them unleashed more than a hundred new features, according to the press materials that accompanied them. There were upgrades for Java support, new animation types, sound plug-ins, e-mail filters, and so on. But not one of these new features—not one—enhanced the basic gesture of clicking on a text link. The very cornerstone of the World Wide Web had been completely ignored under a blizzard of other, gratuitous additions. For those of us who spend a great deal of time "surfing" online, the oversight was maddening. Ask any Web user to recall what first lured him into cyberspace; you're not likely to hear rhapsodic descriptions of a twirling animated graphic or a thin, distorted sound clip. No, the eureka moment for most of us came when we first clicked on a link, and found ourselves jettisoned across the planet. The freedom and immediacy of that movement—shuttling from site to site across the infosphere, following trails of thought wherever they led us—was genuinely unlike anything before it. We'd seen more lively cartoon animations on Saturday-morning television; we'd heard more compelling audio piped out of our home stereos. But nothing could compare to that first link.

What we glimpsed in that first encounter was something profound happening at the level of language. The link is the first significant new form of punctuation to emerge in centuries, but it is only a hint of things to come. Hypertext, in fact, suggests a whole new grammar of possibilities, a new way of writing and telling stories. But to make that new frontier accessible, we need more than one type of link. Microsoft and Netscape may be content with the simple, one-dimensional links of the Web's current incarnation. But for the rest of us, it's like trying to write a novel where the words are separated only by semicolons. (It might make for an intriguing avant-garde experiment, but you're not going to build a new medium out of it.) Fortunately, the world of hypertext has a long history of low-level innovation. More than any other interface element, the link belongs to the cultural peripheries and not to the high-tech conglomerates. Even as the Netscapes of the world ignore hypertext, the novelists and site designers and digital artists are busy conjuring up the new grammar and syntax of linking.

As the word suggests, a link is a way of drawing connections between things, a way of forging semantic relationships. In the terminology of linguistics, the link plays a conjunctive role, binding together disparate ideas in digital prose. This seems self-evident enough, and yet for some reason the critical response to hypertext prose has always fixated on the *disassociative* powers of the link. In the world of hypertext fiction, the emphasis on fragmentation has its merits. But as a general interface convention, the link should usually be understood as a *synthetic* device, a tool that brings multifarious elements together into some kind of orderly unit. In this respect, the most compelling cultural analogy for the hypertext webs of today's interfaces turns out to be not the splintered universe of channel surfing, but rather the damp, fog-shrouded streets of Victorian London, and the mysterious resemblances of Charles Dickens.

•   •   •

"Links of association" was actually a favorite phrase of Dickens. It plays a major role in the narrative of *Great Expectations*—arguably his most intricately plotted work, and the most widely read of his "mature" novels. For Dickens, the link usually takes the form of a passing resemblance, half-glimpsed and then forgotten. Throughout his oeuvre, characters stumble across the faces of strangers and perceive some stray likeness, something felt but impossible to place. These moments are scattered through the novels like hauntings, like half-memories, and it's this ethereal quality that brings them very close to the subjective haze of modernism and the stream of consciousness. Consider Pip's ruminations on his mysterious playmate and love interest Estella: "What was it that was borne in upon my mind when she stood still and looked attentively at me? . . . What was it? . . . As my eyes followed her white hand, again the same dim suggestion that I could not possibly grasp, crossed me. My involuntary start occasioned her to lay her hand upon my arm. Instantly the ghost passed once more and was gone. What *was* it?"

These partial epiphanies are more than just stylistic ornamentation—they serve as the driving force behind the suspense of Dickens's novels. Resolving the half-resemblance, connecting the links, putting a name to the face—these actions invariably give the novel its sense of an ending. They stand for the restoration of a certain orderliness in the face of tremendous disorder. (This is one way in which they mirror the "synthetic" connections of today's hypertext prose.) The "associative links" of the half-glimpsed resemblance are so central to Dickens because they unite his two major thematic obsessions: orphans and inheritances. In the Dickensian novel, the plight of being orphaned at an early age has the same *sine qua non* quality that marital infidelity had in the French novel: you simply can't imagine the form surviving without it. The more complicated novels of the later years— *Bleak House, Our Mutual Friend, Great Expectations*—are teeming with abandoned children, surrogate parents, and anonymous benefactors. The Victorians have a reputation for family-values conservatism, but their most gifted novelist devoted his entire career to dissecting and recombining the family unit, with an inventiveness that would have impressed the Marquis de Sade.

For all the experimentation, of course, Dickens's novels eventually wind their way back to some kind of nuclear family. (It took another twenty years for *that* convention to give way.) And with this "rightful" restoring of the family unit comes another restoration, this one financial. Like almost every other nineteenth-century British novelist, Dickens incessantly structured his narratives around troubled inheritances. There are enough contested wills, anonymous benefactors, and entangled estates in the Victorian novel to keep all the lawyers in Chancery busy for another century altogether. Orphans, of course, made wonderful protagonists for these inheritance plots. In almost every novel, reuniting the dispersed family unit, discovering the links of filiation that connect the main characters—all this is bound up in the rightful disposition of some long-contested estate. What better way to tantalize the reader—haplessly trying to connect those long-separated family lines—than by offering up a suggestive, but unfulfilled resemblance, a hint of filiation. When the moment of realization finally arrives, it has the force of biology *and* capital behind it. It packs a wallop:

I looked at those hands, I looked at those eyes, I looked at that flowing hair, and I compared them with other hands, other eyes, other hair, that I knew of, and with what those might be after twenty years of a brutal husband and a stormy life. . . . I thought how one link of association had helped that identification in the theater and how such a link, wanting before, had been riveted for me now, when I had passed by a chance swift from Estella's name to the fingers with their knitting action and the attentive eyes. And I felt absolutely certain that this woman was Estella's mother.

What makes these links so striking—to the twentieth-century reader, at least—is the fact that they straddle radically different social groups. The family triangle unearthed at the end of *Great Expectations* is that of an escaped convict, a servant, and a young woman of means; in *Bleak House,* it is a baroness, an opium-addicted law stenographer, and an orphan girl brought up by a haute-bourgeois uncle. We know from the outset of each book that the family unit has been dispersed physically; we learn by the end that it has also been separated *economically.* The reconciliation between different social classes has the air of wish fulfillment to it, an imaginary solution—as Freud used to say—to a real contradiction.

There is a strong vein of sentimentality here, of course, but there is also something heroic. Dickens at least attempted to see the "whole" of society in his novels, building a form large enough to connect the lives of street urchins, captains of industry, schoolteachers, circus folk, ladies-in-waiting, convicts, shut-ins, dustheap emperors, aging nobility, and rising young gentlemen. No novelist since has cast such a wide net. No novelist since has dared to try. That is in part because the forces unleashed by the Industrial Revolution had an enormously disassociative power: in the space of twenty or thirty years, they utterly transformed the lives of most British citizens, particularly those residing in the factory towns and the metropolitan center of London. The great burden that Dickens inherited was that of a society in which social roles were no longer clearly defined, where the old codes of primogeniture and noblesse oblige had given way to a dynamic, bewildering new regime, one that seemed to reinvent itself every few years. Certain social critics and historians of the time—most notably Engels and Carlyle—attempted to make sense of this new reality in works of nonfiction; Dickens built his explanatory narratives within the genre of the novel. But the divisions in the society were too broad, too severe, to be broached by ordinary storytelling. To see the relationship between a street orphan and a baroness, you needed a little magic, a little artifice. And so the link of association—leading us inexorably toward a secret history of heritage and inheritance—became the stock device of the Dickensian novel.

When we read the books a century later, the trope can seem forced, almost comical. That the Victorian reading public embraced these fanciful links with such devotion testifies to the divisiveness and the social confusion of the time. The preposterousness of the device suggests just how overwhelming the crisis really was. Dickens's genius—and the key to his popular success—was to understand that a culture so divided against itself could only seek resolution in fairy tales. The "links of association"—all those half-glimpsed resemblances, those partial hauntings—were the building blocks of that fantasy. Their high-tech descendants serve an equivalent purpose today. Where Dickens's narrative links stitched together the torn fabric of industrial society, today's hypertext links attempt the same with information. The imaginative crisis that faces us today is the crisis that comes from having too much information at our fingertips, the near-impossible task of contemplating a colossal web of interconnected computers. The modern interface is a kind of corrective to this multiplying energy, an attempt to subdue all that teeming complexity, make it cohere. And on the World Wide Web, where this imaginative crisis is most sorely felt, it is the link that finally supplies that sense of coherence, like the families reunited at the close of *Bleak House*, or *Hard Times*, or *Great Expectations*. Today's orphans and itinerants are the isolated packets of data strewn across the infosphere. The question is whether it will take another Dickens to bring them all back home again.

As is true for so much in the digital world, the modern practice of linking originates in the creative aftermath of World War II. Not surprisingly, it arose specifically as a response to a perceived crisis of information overload, a crisis set in motion by the extraordinary research explosion of the war years. With so much new data floating around—so many new discoveries, experiments, hypotheses—how were scientists going to make sense of it all?

The question arrives at the very outset of Vannevar Bush's "As We May Think" essay. The problem, as Bush conceived it, was one of discontinuity: our knowledge-creating tools had advanced faster than the knowledge-processing ones. Plenty of information was being generated out there; we just didn't know where to find it. Fifty years before Netscape Navigator, Bush drew on a nautical metaphor to express the thought, already hinting at the provocative idea of information-space: "The summation of human experience is being expanded at a prodigious rate, and the means we use for threading through the consequent maze to the momentarily important item is the same as was used in the days of square-rigged ships." As a corrective to this plight, Bush proposed an information speedboat of sorts, a device that was half microfilm machine and half computer. He called it the Memex.

It consists of a desk, and while it can presumably be operated from a distance, it is primarily the piece of furniture at which [the user] works. On the top are slanting translucent screens, on which material can be projected for convenient reading. There is a keyboard, and sets of buttons and levers. Otherwise it looks like an ordinary desk.

In one end is the stored material. The matter of bulk is well taken care of by improved microfilm. Only a small part of the interior of the memex is devoted to storage, the rest to mechanism. Yet if the user inserted 5000 pages of material a day it would take him hundreds of years to fill the repository, so he can be profligate and enter material freely.

Most of the memex contents are purchased on microfilm ready for insertion. Books of all sorts, pictures, current periodicals, newspapers, are thus obtained and dropped into place. Business correspondence takes the same path. And there is provision for direct entry.

We will return to Bush's mechanical blueprint for the Memex in the concluding chapter, but for now let us consider his navigational device. Information *storage*, after all, isn't the problem here, as Bush pointed out in his opening argument. We have plenty of "summation" lying around, whether it's on our desktop or in the local library; what we don't have is a way of "threading" through all that data. Bush's proposed solution should probably go down in history as the birth of hypertext, at least in its modern incarnation. Only he chose to imagine the "links of association" connecting all that data as "trails," not links. At one point, he even refers to experienced Memex users as trailblazers—a term that would have fitted well with the "new frontier" rhetoric of recent cyber-boosterism. It certainly would have been an improvement on the couch-potato passivity of the "surfing" argot.

At first glance, trails appear to have much in common with the modern link; they serve as a kind of connective tissue, an information artery, that threads together documents with some shared semantic quality. Trails, in other words, are a way of organizing information that doesn't follow the strict, inflexible dictates of the Dewey decimal system or other hierarchical conventions. Documents can be connected for more elusive, transient reasons, and each text can have many trails leading to it. Our traditional ways of organizing things—library books, say, or physical elements—are built around fixed, stable identities: each document belongs to a specific category, just as each element has a single block on the periodic table. Bush's system was closer to those half-resemblances of Dickens's novels: links of association, tantalizing, but not fully formed.

This implied a profound shift in the way we grapple with information. The previous century had been dominated by the encyclopedic mentality (famously parodied in Flaubert's slapstick novel *Bouvard et Pécuchet*) in which the primary goal of information management was to find the proper slot for each data package. Bush turned that paradigm on its head. What made a nugget of information valuable, he suggested, was not the overarching class or species that it belonged to, but rather the *connections* it had to other data. The Memex wouldn't see the world as a librarian does, as an endless series of items to be filed away on the proper shelf. It would see the world the way a *poet* does: a world teeming with associations, minglings, continuities. And the trails would keep that radiant universe bound together.

What Bush described was essentially a literary view of the world, one probably best realized in Bloom's rambling internal monologue in *Ulysses*, and in the associative free-for-all of most surrealist writing. Recent advances in neuroscience suggest, though, that Bush's connective model may be a mechanical analog of the way the brain works: an intricate assemblage of neurons connected by trails of electrical energy, generating information out of connections rather than fixed identity. It's not as if the brain reserves a specific chunk of physical real estate for the idea of "dog" and another for "cat." The ideas emerge out of thousands of separate neurons firing, in combinations that reorganize themselves with each subtle shift in meaning. The connections between those neurons create the thought; the individual neurons are just building blocks.

Part of Bush's vision for the Memex looks uncannily like our present-day experience of the Web, with a predictably heavy emphasis on the research benefits promised by the new technology, and little attention paid to more, shall we say, *recreational* pursuits. (You get the sense that Dr. Bush would have had a hard time adjusting to the "trails" connecting the subcultures of online porn.) Sections of his essay read like a run-of-the-mill, late-nineties AT&T ad, or an overenthused product review in *Wired:*

> The patent attorney has on call the millions of issued patents, with familiar trails to every point of his client's interest. The physician, puzzled by his patient's reactions, strikes the trail established in studying an earlier similar case, and runs rapidly through analogous case histories, with side references to the classics for the pertinent anatomy and histology. The chemist, struggling with the synthesis of an organic compound, has all the chemical literature before him in his laboratory, with trails following the analogies of compounds, and side trails to their physical and chemical behavior.

The scenarios sound like a feasible description of today's online databases and vertical-market CD-ROMs—though as always, the promise of "information at your fingertips" works better on paper than it does in real life. (The gap is probably forgivable in this one instance, considering that the Memex itself was the ultimate in vaporware.) But if part of Bush's vision anticipates the present-day shape of the World Wide Web, another part greatly exceeds it. Despite the fury of innovation and the massive R&D expenditures of the past decades, one of the Memex's essential features remains off-limits to most contemporary Web browsers. Consider this description:

The owner of the memex, let us say, is interested in the origin and properties of the bow and arrow. . . . He has dozens of possibly pertinent books and articles in his memex. First he runs through an encyclopedia, finds an interesting but sketchy article, leaves it projected. Next, in a history, he finds another pertinent item, and ties the two together. Thus he goes, building a trail of many items. Occasionally he inserts a comment of his own, either linking it into the main trail or joining it by a side trail to a particular item. When it becomes evident that the elastic properties of available materials had a great deal to do with the bow, he branches off on a side trail which takes him through textbooks on elasticity and tables of physical constants. He inserts a page of longhand analysis of his own. Thus he builds a trail of his interest through the maze of materials available to him.

Anyone who has spent any time roaming across the Internet will immediately recognize the difference here. Bush's Memex owner *builds* that "trail of interest" as he explores the information-space on his desk. Surfers, as a rule, *follow* trails of interest, through links that have been assembled in advance by other folks: designers, writers, editors, and so on. The Web surfer depends on the charity of others for his associative links; the "trailblazer" rolls his own. And most important, the trails endure. They remain part of the Memex's documentary record; the connection between the bow and the principles of elasticity isn't simply strung together momentarily, only to be discarded hours later. The connection remains permanently etched onto the Memex's file system. Five years after this initial research, a return to the material on elastics might send our Memexer off to the bow-and-arrows article, or deliver up his long-forgotten notes on the subject. That accumulated record of past trails means that the device grows smarter—or at least more associative—the more you use it, as the file system is laced together by thousands of associative trails.

Your average Nethead can create bookmarks, of course, but these are just momentary excerpts from a longer train of thought, like snapshots or postcards mailed home from an overseas vacation. The journey itself—the movement from thought to thought, document to document—is the key here. Bookmarking a single page barely scratches the surface. Most of us carry around bookmark files littered with random sightings, recommendations, favorite locales, secret hideaways, and so on. It's a remarkably personal, idiosyncratic list. (Trading bookmark files—one of the first rituals to develop in Web culture—has a wonderfully confessional quality to it, like letting someone eavesdrop on your therapy sessions.) But despite their personalized texture, those bookmarks have no connection to one another. They're isolate units, monads. You can create a master list of all your favorite resources, but there's no way to describe the relationships between them, the links of association that make that personal web intelligible to you.

The Memex was designed to organize information in the most intuitive way possible, based not on file cabinets or superhighways but on our usual habits of thinking—following leads, making connections, building trails of thought. Bush wanted the Memex to respond to the user's worldview; the trails would wind their way through documents in varied, idiosyncratic ways, threading through the information-space at the user's discretion. No two trails would be exactly alike. The Web has realized much of Bush's vision, but the core insight—the need for a trail-building device— remains unfulfilled, at least on the Internet. (Several group- ware products—Lotus Notes, for example—have come close to the Memex's trail-building technology.) Most Web browsers still dutifully follow the links that are served up to them, without any means of creating their own associative trails in return. The Web should be a way of seeing new relationships, connecting things that might have otherwise been kept sepa- rate. Clicking on other people's links may be less passive than the old, sedentary habit of channel surfing, but until users can create their own threads of association, there will be few gen- uine trailblazers on the Net.

The irony here, of course, is that a middle-aged army scientist, writing thirty years before the first PC, under- stood interactivity better than all the Web titans in Silicon Valley. Perhaps this shouldn't come as a surprise. After all, sometimes the best way to understand a technology is to approach it with no expectations, no preconceived ideas. Unhampered by any historical precedent, Bush was free to con- jure up a device for "augmenting" thought based on his own flights of fancy. Today's technologists may be too trapped within the "surfing" paradigm—clicking absentmindedly on links supplied by others—to recognize the value of being able to link back, to blaze your own trail through information-space.

We may be anesthetized to it now, but the truth is, clicking on links once had a certain air of sedition to it, back in the early days of hypertext, before the overnight success of the World

Wide Web. Hypertext first captured the public's imagination as a literary genre, most famously in Michael Joyce's 1993 work, *Afternoon, A Story*. The modest title concealed a labyrinth of narrative passageways, winding across one another, or snaking back to their origins. (Hypertext writers tend to play up their prepositions a great deal; the literature abounds with quirky, precious phrases such as "reading through and over the text.") Like most interface advances, this new form had a politics to it, though in this case the rhetoric of liberation was a little closer to the surface. Hypertext advocates drew on a tradition that dated back to the literary theorists of the sixties, to essays like Roland Barthes's influential "Death of the Author." The Parisian *philosophes* of '68 had called for a revolution in reading habits, a kind of grassroots aestheticism wherein the *reader* shapes the experience of a text more than the author does.

In its original iteration, this "reader's revolt" was mostly a figure of speech, and a self-aggrandizing one at that. In its most elemental form, it argued that the critic—and not the author—had the final say in what a book "meant," and because there were many critics out there, each lugging along a competing interpretation, it was unlikely that the book would ever arrive at a unified, stable meaning. There was some truth to be found amid all the manifestos and tortured syntax. (Surely all great works of art possess multiple levels of meaning, levels that are brought out by the aptitudes and inclinations of the audiences that receive them.) But all too often, the "death of the author" came across as a case of self-interested ressentiment. "Reader centrism" translated into "critic centrism," which translated into tenured faculty positions and high-priced lecture tours.

Hypertext fiction could go beyond all that, its advocates argued. It would literalize the metaphor of the "reader's revolt." Hypertext would be a more egalitarian form, where the reader would create the narrative by clicking on links and following different story lines, like the old "choose your own adventure" children's books. The work itself would be less like a narrative in the strict sense of the word, and more like an environment. (It was no accident that the first authoring software for hypertext fiction was called Story-Space.) *Afternoon, A Story* was widely hyped as a harbinger of this great textual revolution, and it was scrutinized like tea leaves or sheep entrails for signs of things to come.

As it turned out, *Afternoon* didn't make for a particularly good case study. The writing itself was fairly experimental; you got the sense that it would have remained disconcertingly nonlinear had it been published in traditional book form. It was a bit like watching a Godard film with a projectionist who insists on randomly swapping reels. The links between different pages seemed more anarchic than free-associative; the idea of an overarching narrative dropped out altogether, and what you were left with resembled a collection of aphorisms more than anything else. You couldn't help wondering how a John Grisham novel would fare in the medium, where a strong narrative might make your reading "choices" more consequential.

One casualty of hypertext (at least in Joyce's hands) was that old, time-honored sense of an ending. I had assumed from the outset that *Afternoon* would offer many potential endings, but a few readings of the story left me with the sense that "closure"—as Joyce calls it in the introduction— had been abandoned wholesale. "When the story no longer progresses," he wrote, "or when it cycles, or when you tire of the paths, the experience of reading it ends." This is closure as entropy rather than resolution; the story stops when it bores you. Hypertext advocates saw this as another way of empowering the reader at the author's expense, though that line always sounded a little suspicious to me. I remember as a teenager "ending" *Crime and Punishment* halfway through the novel precisely because I had "tired of its paths." Had I known then about the politics of hypertext, I might never have made it past the first chapter.

There is another limitation to hypertext fiction's liberation theology. The politics of reading, after all, aren't simply a matter of a confrontation between author and reader. There's also that other, crucial dimension: the readers' shared experience, the broader social bond that develops out of having read the same narratives. This shared experience was an essential component of Dickens's success as a novelist. The links of association bound together not only the disparate social universes of his characters, all those ruffians, ladies-in-waiting, stockbrokers, and day laborers. They also bound together a nation of readers. Without that collective resonance, the faint, reassuring rustle of a thousand fingers turning the same pages in unison, Dickens's imaginary resolutions would have lost their force. And herein lies the great distinction between the Dickensian link and its hypertext descendants. Dickens's links worked in the service of unification, welding together the fictional lives of his characters as well as the imaginations of his readers. Joyce's links proceed in the opposite direction. They fragment the reading experience, scatter it into hundreds of variations, to the point where every reading conjures up a different story.

There's something thrilling about that new open-endedness, but also something profoundly lonely. After I fin-

interface culture

ished with *Afternoon,* I rang up a few friends who had also meandered through it, looking for feedback. In each phone call, we talked excitedly for a minute or two about the medium and its possibilities, but the second we turned to the content of the story, the conversation grew stilted and uneven. We were talking, it turned out, about very different stories. Each reading had produced an individual, private experience. At these moments, struggling for common ground over the telephone, hypertext felt less like an exercise in literary democracy and more like an isolation booth.

Although the hypertext soothsayers were right to sense something significant brewing in the new grammar of links, most of them were thinking on the wrong scale. Hypertext was supposed to revolutionize the way we tell stories, but it ended up transforming our *sentences* instead. Nowhere is this more apparent than in the World Wide Web itself—now the great breeding ground for hypertext innovation. The Web first drifted into prominence near the end of 1994, just as the public fascination with nonlinear fiction was hitting a high point. Joyce had been named to *Newsweek's* list of digital savants; the *New York Times Book Review* had run several extended essays on hypertext novels, laced with the obligatory references to Cortázar and Calvino; Sven Birkerts had published his assault on the forking paths of nonlinear narrative, *The Gutenberg Elegies.* The Web was seen as a logical continuation of this trend: a global medium for hypertext narrative. Soon we'd all be navigating through elaborate storyspaces on our desktop PCs, stitching our own tailor-made plots together with each mouse click. Journalists would file stories in a more three-dimensional format—as an array of possible combinations rather than a unified piece.

The links would transform our most fundamental expectations about traditional narrative. We'd come to value environment over argument, shape-shifting over consistency.

But looking back now, after a few years of press releases and vaporware, what strikes you is how little of this came to pass. The great preponderance of Web-based writing is unapologetically linear. Almost all journalistic stories are single, one-dimensional pieces, articles that would be exactly the same were they built out of ink and paper instead of zeros and ones. (Many of them, of course, are simply digital versions of print originals.) If there is reader-centric navigation, it comes from hopping from article to article and from site to site. The individual articles themselves rarely offer any navigational options at all. Links do appear in some articles, but they're usually pointing to the Web sites of companies that happen to be mentioned in the piece—yet another way of accentuating brand identity, like a registered trademark or a logo. This is a particularly mindless use of hypertext. Illuminating a passing reference to Apple Computer with a link to "www.apple.com" might create the appearance of hypertext prose, but in actuality it's gratuitous, yet another case of digital window dressing. Finding a corporate Web site is one of the easiest tasks on the Web: it usually involves tacking a ".com" suffix onto the company's name and punching that into your browser. Reading an article about Apple Computer doesn't make you want to check out its home page; it makes you want to read other, related articles on the same topic, or zoom in on one particularly tantalizing idea, or click over to a reader discussion about the company's future.

To be fair, a handful of Web publishers have integrated "related reading" pointers into their articles, though there is a strange compulsion to keep those links separated from the primary text. (Slate, for instance, trots out its links at the end of each article.) Other, community-driven sites—like HotWired and Electric Minds—feature excerpts from reader commentary in the margins of the top-level articles. But even the more adventurous, envelope-pushing sites like Word seem more preoccupied with multimedia frills than with associative links. When Stefanie Syman and I first designed FEED, we included two sections—Document and Dialog—that relied extensively on the new dimensions of hypertext. Document allowed readers and contributors to attach their own commentary to a primary text, like birds perched on the backs of lumbering elephants. Dialog deposited a panel of critics in a "conversation space," where each written remark led off in several directions; one sentence might generate a string of rebuttals and counterrebuttals, while another sentence might lead off to a mild clarification from the original author, or to a missive sent in by a reader. You didn't read so much as *explore* the Dialog, and like most interesting spaces, you'd stumble across a new passageway every time you went back to it. This was journalism for trailblazers, we thought, and we assumed that other Web publications would soon adopt similar journalistic storyspaces.

129

•

l
i
n
k
s

But two years later, the FEED Dialog remains one of the Web's most complex hypertext environments, at least among the mainstream publications. (More serpentine structures have been built on the margins of avant-garde fiction by hypertext trailblazers like Carolyn Guyer and Mark Amerika.) It may be that readers genuinely prefer the ordered, author-centric direction of traditional storytelling, and so more complex structures will remain the exception to the rule. But my hunch is that the appetite for nonlinear prose will grow as we acclimate ourselves to these new environments—and to the strange new habits of reading that they require. Here again the legacy of channel surfing has done the Web a great disservice. The metaphor suggests a certain agitated indifference, zapping randomly from source to source. But moving through a hypertext space, following links of association, is an intensely focused activity. Channel surfing is all about the thrill of surfaces. Web surfing is about depth, about wanting to know *more*. But if you can't see that distinction, if you imagine the mouse as the poor cousin of the remote control, then of course you're not going to create documents that fully exploit the power of hypertext. There's plenty of programming designed for trigger-happy surfers on MTV; why bother lowering yourself to that common denominator on the Web?

Fortunately, ill-advised metaphors can't possibly curtail *all* innovation, particularly with a medium as democratic as the Web. As it turns out, the most interesting advances have taken place on the micro level of syntax, rather than the macro level of storytelling. This is one of those wonderful occasions—frequent in high-tech history—when the pundits and trendspotters have us looking in one direction and the exciting stuff ends up happening somewhere else. Hypertext links were supposed to be a storytelling device, but their most intriguing use has proved to be more syntactical, closer to the way we use adjectives and adverbs in our written language. The link was going to engender a whole new way of telling stories. It turned out to be an element of style.

Nowhere is this more apparent than in the irony-drenched column of Suck. Launched anonymously by a pair of Unix hackers in the HotWired basement, Suck is now generally regarded as the ultimate do-it-yourself, self-publishing success story in the Web's short history. The daily column took aim at the Web's relentless march toward the commercial mainstream (this was still news at the time), riffing caustically on the bloated, straining-to-be-visionary pronouncements of the "digital elite" or the inane online brochures of most corporate Web sites. The Sucksters liked to play themselves off as slackers and malcontents, lacing their columns with crack-smoking jokes and references to their being "postliterate." But the bad-boy posturing couldn't mask the intelligence and inventiveness of the prose, with its elliptical phrasing and penchant for extended metaphors. This onscreen style was both a curse and a blessing. The columns invariably sounded wonderful on first reading, but the layering of rhetoric made it difficult to pin down exactly what it was they were saying. Despite all the Budweiser jokes, what came to mind reading Suck was the cagey, intricate language of literary theory, the willed evasiveness of someone trying to use language to talk about how language doesn't work.

For a long time, I was puzzled by my return visits to Suck. Too many times the prose had seemed deliberately obscure, as if it were actively trying to repel its audience, inundating them with in-jokes, pop-culture references, French theory, and bathroom humor. Certain sentences had a kind of elusive, shimmering quality to them, as if you were seeing them at a great distance. You sensed that a tangible meaning lurked in the mix—if only you had the time to disentangle all the subsidiary clauses, parse out all the throwaway references. Consider this obtuse, but representative, example:

> *In the new infomockracy, the cafe*
> *tables have been overturned. The*
> *stiffs chained to hollowed-out*

terminals are now on the
bleeding edge, while the most
observed of old-line cultural
observers are merely blunted. No
more reheeling your Manolo mules
every three weeks—a lack of
mobility confers an advantage.
Though boxed into cubicles, the
new counterparts of Whether
Overground footsoldiers have
freer, faster access to the
entrails and tea leaves of
hipster life.

Normally, of course, I'd have little patience with the onslaught of mixed-metaphor allusions, particularly on a Web site. But I found myself returning to the Sucksters, not so much to read what they had to say, but to figure out how they were saying it. After a few weeks of study, I began to realize that the uncanny, ethereal quality of the prose—particularly uncanny given the earthiness of the words themselves—was a side effect of the links. Like the passing resemblances of *Great Expectations,* the links triggered that sense of mystery, the sense of a code half-deciphered.

Suck's great rhetorical sleight of hand was this: whereas every other Web site conceived hypertext as a way of *augmenting* the reading experience, Suck saw it as an opportunity to withhold information, to keep the reader at bay. Even the sophisticated Web auteurs offered up their links the way a waiter offers up fresh-ground pepper: as a supplement to the main course, a spice. (Want more? Just click here.) The articles themselves were unaffected by the "further readings" they pointed to. The links were just addenda, extensions of the primary argument. The Sucksters took the opposite tack. They used hypertext to condense their prose, not expand it. The benefits were clear: they could move faster through their sentences if they linked out strategically to other documents. They didn't need to spell out their allusions; they could just *point* to them and leave it up to the reader to follow along. So they left things out, and let the trails do the work. They buried their links mid-sentence, like riddles, like clues. You had to trek out after them to make the sentence cohere.

The rest of the Web saw hypertext as an electrified table of contents, or a supply of steroid-addled footnotes. The Sucksters saw it as a way of phrasing a thought. They stitched links into the fabric of their sentence, like an adjective vamping up a noun, or a parenthetical clause that conveys a sense of unease with the main premise of the sentence. They didn't bother with the usual conventions of "further reading"; they weren't linking to the interactive discussions among their readers; and they certainly weren't building hypertext "environments." (Each Suck article took the resolutely one-dimensional form of a thin column snaking down an austere white page.) Instead, they used links like modifiers, like punctuation—something hardwired into the sentence itself. Most hypertext follows a centrifugal path, forcing its readers outward. The links encourage you to go somewhere else. They say, in effect: When you're done with this piece, you might want to check out these other sites. More sophisticated hypertext story-spaces say: Now that you've enjoyed this particular block of text, where would you like to go next? Suck, on the other hand, pointed its readers outward only to pull them back in, like

Pacino's tragic dance with the Mob in the *Godfather* trilogy. The links were a way of cracking the code of the sentences; the more you knew about the site on the other end of the link, the more meaningful the sentence became.

In its simplest form, Suck's hyperlinks worked the way "scarequotes" work in slacker idiom. They labored in the service of irony, undermining the seriousness of the statement, like a defense mechanism or a nervous twitch. Suck was notorious for linking to *itself* at any mention of crass commercialism or degeneracy. You'd see *sellout* or *jaded* highlighted in electric blue, and you'd click dutifully on the link—only to find yourself dropped back into the very page you were originally reading. The first time it happened, you were likely to think it was a mistake, a programmer's error. But after a while, the significance of the device sank in. By linking to itself, Suck broke with the traditional, outer-directed conventions of hypertext: what made the link interesting was not the information at the other end—there was no "other end"—but rather the way the link insinuated itself into the sentence. Modifying "sellout" with a link back to themselves was shorthand for "we know we're just as guilty of commercialism as the next guy"—in the same way that scarequotes around a word is shorthand for "I'm using this term but I don't really believe in it." The link added another *dimension* to the language, but not in the storyspace sense of the word. You never felt that you were exploring a Suck piece or navigating through an environment. You were just reading, but the sentences that scrolled down the screen had a strange vitality to them. They were more resonant somehow, and the hypertext shorthand allowed them to do much more with less.

The self-referential links were actually the easiest codes to decipher. Other combinations took more effort.

Consider this sentence from an end-of-the-year column. It read, at first, like a seasonal good tidings from one Web publication to another, but once you unraveled all the links, the words took on a darker, more cutting tone. "We are pleased to see that FEED is still worth the effort, though occasionally extraneous." It's an intelligible enough phrase, if a little vague. But reading the sentence through the lens of hypertext sharpened the image noticeably. The word *effort* pointed to an article we had run at FEED critiquing the WebTV product by Sony and Philips; the word *occasionally* linked to a Suck piece, penned months earlier, on the same topic. *Extraneous* pointed to another Suck article that predated ours—this one less critical of WebTV. When you added it all up, the "meaning" of the sentence was a good deal more complicated than the original formulation. Like one of Freud's dream studies, the sentence had a manifest and a latent content. The former was clear-cut, straightforward: "We are pleased to see that FEED is still worth the effort, though occasionally extraneous." The latter was more oblique, something like: "We're still fans of FEED, though they tend to be about two months behind us, and they tend to rip off our ideas when they finally catch up—like this WebTV travesty." As in the dream work of psychoanalysis, the latent content had a way of infecting the manifest content. After you deciphered the links, the phrase *worth the effort* began to sound more and more derogatory, as though the readers were laboring under the "effort" rather than being rewarded for it.

It may sound like an unlikely comparison, given Suck's postliterate pretensions, but what these passages remind me of are the famous lines from Wallace Stevens's "Thirteen Ways of Looking at a Blackbird":

I do not know which to prefer,
The beauty of inflections
Or the beauty of innuendoes.

Stevens describes the gap between literal language ("the beauty of inflections") and those subtle but still meaningful silences *between* words, their resonance, their beautiful innuendo. Suck's hypertext links seem to me to straddle that gap. They hover over the language, shadow it, part inflection and part innuendo. Who can say where the literal meaning lies? You can read the sentence straight, ignoring the links altogether, and it will indeed make sense, though you can't help but feel that something has been lost in the translation. But it's just as hard to imagine the links as an integral part of the sentence's meaning, as integral as the words themselves. Wouldn't that be a whole new way of writing? And even if we *are* witnessing the birth of a new type of language, surely it's not the offspring of a bunch of postliterate hackers?

Suck's use of hypertext is actually a bit less momentous in its implications, and a bit more encouraging. Making sense of those links brings us full circle, all the way back to the restricting language of Web surfing. What you can see in Suck's oblique syntax is not the birth of a new language, but rather the birth of a new type of *slang*. It's a jargon, but it's not built out of words or phrases. It's the slang of associations, of relationships between words. The slang evolves out of the way you string together information, the way you make your references, and not the words you use. If punctuation can become an element of slang (think scarequotes again), then why not links?

My guess is that old Vannevar Bush would have been delighted with the layered, associative syntax of Suck—despite the slacker invectives. Nothing could be healthier for the future of hypertext than a bunch of kids wrestling around with new intonations, new twists on old habits. That's what keeps language moving, after all—whether it's oral, print, or digital. The Netscapes and Microsofts of the world may ignore hypertext for years to come, but as long as the forces of popular idiom keep churning out the innovations, the dream of the Memex will continue to grow more vivid, more lifelike.

And yet for all their significance, links are not the only linguistic component of the modern interface. The demise of the command-line regime may have dealt a mortal blow to the supremacy of text over image in interface design, but simple words still play an enormous role in the contemporary interface. If anything, that role looks to become more critical to our information-spaces in the next decade, for reasons that are only now becoming apparent. The next chapter makes the case for the renewed importance of text in future interface designs, but it begins with the digital revolution's most influential gift to written language so far: the word processor.

TEXT

I was twelve when my parents shelled out for our first home PC—an Apple IIe souped up with an astonishing 32K of RAM—and while my recollections of the preceding years are not particularly vivid, I can still conjure up a little of the rhythm of life back then, in the dark ages before the digital revolution. Among my peers, this sometimes seems to be an unusual ability. I often hear friends wonder aloud: "How did we ever get along without e-mail and word processors?" And yet for the most part I can readily imagine how things happened in that world, the pace of that more settled and disconnected existence. It all seems rather obvious to me. We got along because we didn't know what we were missing. Folks have always griped about the postal service's sluggish performance, but the lag time only becomes intolerable once you have a taste of e-mail.

It's not life without computers that confounds me; it's life in the strange interregnum after the PC first appeared in our house. I lived within thirty feet of a fully functional computer from my twelfth to my eighteenth year, and yet the sad truth is, I used it almost exclusively as a presentation device for those six years, like a visit to Kinko's at the end of a term paper. Whatever I was writing—papers, poems, stories, plays—I dutifully etched out by hand on yellow legal pads, crossing out passages, scribbling new lines in the margins. Only when the language had reached a tolerable state did I bother typing it into the PC. The idea of composing on (and not transcribing into) the machine seemed somehow inauthentic to me. It was more like typing than writing, in Truman Capote's memorable phrase, somehow more mechanical, more mediated, a few steps removed from the whole books-on-tape phenomenon.

This continues to be a normal state of mind for the millions of people who still sense something menacing in the glare of the PC monitor, who find themselves more perplexed than enlightened by the digital revolution. But that fifteen-year-old version of me didn't belong to that demographic: I genuinely liked computers, and spent the requisite hours of my adolescence frittering away my allowance at the arcade. Like many kids of my generation, I dabbled with rudimentary programming languages (BASIC and Pascal) long enough to toss a few stray colored pixels up on the screen or scroll through the call-and-response formula of the medieval text adventures then in vogue. I wasn't even close to being what we would now call a "hacker," but I certainly felt confident enough with the PC to put a new word processor through its paces without spending much time with the manual. (In those days, you had to give the documentation at least a cursory glance before booting up the software.) I harbored no ill will toward the machine, no superstitions. But I could not bring myself to write on it.

Fast-forward a decade or two, and I can't imagine writing *without* a computer. Even jotting down a note with pen and paper feels strained, like a paraplegic suddenly granted the use of his legs. I have to *think* about writing, think about it consciously as my hand scratches out the words on the page, think about the act itself. There is none of the easy flow of the word processor, just a kind of drudgery, running against the thick grain of habit. Pen and paper feel profoundly different to me now—they have the air of an inferior technology about them, the sort of contraption well suited for jotting down a phone number, but not much beyond that. Writing an entire book by hand strikes me as being a little like filming *Citizen Kane* with a camcorder. You can make a go at it, of course, but on some fundamental level you've misjudged the appropriate scale of the technology you're using. It sounds appalling, I know, but there it is. I'm a typer, not a writer. Even my handwriting is disintegrating, becoming less and less *my* handwriting, and more the erratic, anonymous scrawl of someone learning to write for the first time.

I accept this condition gladly, and at the same time I can recall the predigital years of my childhood, writing stories by hand into loose-leaf notebooks, practicing my cursive strokes and then surveying the loops and descenders, seeing something there that looked like me, my sense of selfhood scrawled onto the page. On a certain level these two mental states are totally incompatible—bits versus atoms—but the truth is I have no trouble reconciling them. My "written" self has always fed back powerfully into my normal, walking-around-doing-more-or-less-nothing self. When I was young that circuit was completed by tools of ink and paper; today it belongs to the zeros and ones. The basic shape of the circuit is unchanged.

But what interests me now, looking back on it, is the *transition* from one to the other. That feeling of artificiality that undermined me as I typed into a word processor, the strangeness of the activity—all this is very difficult to bring back. How could I have resisted so long? Sure, the software was less powerful back then, but the basic components of word processing—the cutting and pasting, the experimentation, the speed of typing—were all very much in place. There were clear advantages to working on the computer, advantages I genuinely understood and appreciated. But they were not compelling enough to dissipate the aura of inauthenticity that surrounded the machine. My writing didn't seem real on the screen somehow. It felt like a bureaucratic parody of me, several steps removed, like a recycled Xerox image shuffled around the office one too many times.

So now I wonder: what force finally brought me over to the other side? After more than half a decade of tinkering with computers, what was it that finally allowed me to recognize myself in those bright pixels on the screen, to see those letterforms as real extensions of my thought? I wasn't totally aware of it at the time, of course, but I can now see that what drew me into the language-space on the screen was nothing less than interface design. The Mac's paper-on-desktop metaphor—the white backdrop, the typographic controls, Alan Kay's stacked windows—lured me away from real-world paper. The "user illusion" sucked me in, and I was hooked forever. I'd understood the benefits of using a word processor before I bought my Mac, but it took a fully realized graphic interface to make me feel comfortable enough to use one for honest-to-God *writing*. Everything before that twelve-point New York font first appeared on the screen, black pixels marching boldly across the whiteness—everything before that was just transcribing.

I suspect there are millions of people with similar stories to tell: the mind naturally resists the dull glare of the screen, feels ill at ease with it, unnatural. And then something in the user experience changes—the "direct manipulation" of the mouse, perhaps, or the resolution of the display—and suddenly you find yourself at home in front of the machine, so acclimated to the environment that you're no longer fighting the software. Before you know it, you're composing directly into the word processor, and the artifice, that original sense of mediation, is gone.

There are two lessons here, one relatively straightforward, the other more indirect. It's clear that the graphic interface played a crucial role in creating today's colossal market for word-processing applications, a market drawn not only to the functionality of the products but also to their look-and-feel. Plenty of us labored along with word processors in the days of the command-line interface, but the ease and fluidity of today's digital writing owe a great deal to the aesthetic innovations of the desktop metaphor. It's not just that the software has accumulated more features. It's also that the software has grown more seductive, more visually appealing over that period. For the creative mind, wrestling with language on the screen, that heightened visual sensibility can be enormously comforting.

But this is more than just a story about the sales records set by WordPerfect and Microsoft Word in the past decade. It also extends beyond the long-term trend of folks becoming more comfortable with their word processors as the user interface grows increasingly sophisticated. The truly interesting thing here is that using a word processor changes how we write—not just because we're relying on new tools to get the job done, but also because the computer fundamentally transforms the way we conjure up our sentences, the thought process that runs alongside the writing process. You can see this transformation at work on a number of levels. The most basic is one of sheer volume: the speed of digital composition—not to mention the undo commands and the spell checker—makes it a great deal easier to churn out ten pages where we might once have scratched out five using pen and paper (or a Smith-Corona). The perishability of certain digital formats—e-mail being the most obvious example—has also created a more casual, almost conversational writing style, a fusion of written letter and telephone-speak.

But for me, the most intriguing side effect of the word processor lies in the changed relationship between a sentence in its conceptual form and its physical translation onto the page or the screen. In the years when I still wrote using pen and paper or a typewriter, I almost invariably worked out each sentence in my head before I began transcribing it on the page. There was a clear before and after to the process: I would work out the subject and verb, modifiers, subsidiary clauses in advance; I would tinker with the arrangement for a minute or two; and when the mix seemed right, I'd turn back to the yellow legal pad. The method made sense, given the tools I was using—changing the sequence of words after you'd scrawled them out quickly made a mess of your document. (You could swap phrases in and out with arrows and cross-outs, of course, but it made reading over the text extremely unpleasant.) All this changed after the siren song of the Mac's interface lured me into writing directly at the computer. I began with my familiar start-and-stop routine, dutifully thinking up the sentence before typing it out, but it soon became clear that the word processor eliminated the penalty that revisions normally exacted. If the phrasing wasn't quite right, you could rearrange words with a few quick mouse gestures, and the magical "delete" key was always a split second away. After a few months, I noticed a qualitative shift in the way I worked with sentences: the thinking and the typing processes began to overlap. A phrase would come into my head—a sentence fragment, an opening clause, a parenthetical remark—and before I had time to mull it over, the words would be up on the screen. Only then would I start fishing around for a verb, or a prepositional phrase to close out the sentence. Most sentences would unfold through a kind of staggered trial and error—darting back and forth between several different iterations until I arrived at something that seemed to work.

It was a subtle change, but a profound one nonetheless. The fundamental units of my writing had mutated under the spell of the word processor: I had begun by working with blocks of complete sentences, but by the end I was thinking in smaller blocks, in units of discrete phrases. This, of course, had an enormous effect on the types of sentences I ended up writing. The older procedure imposed a kind of upward ceiling on the sentence's complexity: you had to be able to hold the entire sequence of words in your head, which meant that the mind naturally gravitated to simpler, more direct syntax. Too many subsidiary clauses and you lost track. But the word processor allowed me to zoom in on smaller clusters of words and build out from there—I could always add another aside, some more descriptive frippery, because the overall shape of the sentence was never in question. If I lost track of the subject-verb agreement, I could always go back and adjust it. And so my sentences swelled out enormously, like a small village besieged by new immigrants. They were ringed by countless peripheral thoughts and show-off allusions, paved by endless qualifications and false starts. It didn't help matters that I happened to be under the sway of French semiotic theory at the time, but I know those sentences would have been almost impossible to execute had I been scribbling them out on my old legal pads. The computer had not only made it easier for me to write; it had also changed the very substance of what I was writing, and in that sense, I suspect, it had an enormous effect on my thinking as well.

But if the early days of the graphic interface altered the way we put our words together, the textual dimension of interface design has been sorely neglected in recent years, as if all the potential variations of linguistic manipulation had already been explored. As the history of the word processor suggests, though, the translation of text into digital form can produce extraordinary—and unpredictable—secondary effects. The contemporary interface climate may appear bleak when it comes to the role of words, but as with so much in the high-tech world, it doesn't pay to judge by appearances. We may, in fact, be on the cusp of a textual paradigm shift as profound as the one ushered in by the rise of the word processor. All the elements are in place for such a revolution; we just need the breakthrough software to bind the elements together in a coherent whole.

This shortsightedness is a familiar story, of course. New technologies are always misunderstood at their birth, often by the people closest to them. From 1877 to 1881, Thomas Edison managed to conjure up close to half of the twentieth century's major inventions, but some of the most entertaining—and illustrative—stories from his Menlo Park years involve his blunders and misconceptions as much as his genius. The phonograph is a wonderful case in point. Edison first envisioned the device as a means of storing recorded telephone conversations, an audio archive that would give substance to those otherwise fleeting, ethereal exchanges over the wire. The disembodied, nowhere-land quality of the telephone had disturbed its first generation of users, and Edison had been tinkering with ways to give it a material grounding, like the printed transcripts of telegraph sessions. And so he came up with the phonograph as a sidekick of sorts, an accomplice—like an answering machine that keeps on recording after you pick up the line.

With the benefit of hindsight, of course, we can see that Edison had to make *two* mistakes to arrive at that assumption: he had to overlook the phonograph's capacity for playing back recorded music, and he had to assume—wrongly, as it turned out—that phone conversations would be regularly recorded. These were hardly rounding errors in Edison's calculations; they involved the most fundamental properties of each medium. In one case, Edison mistakenly assumed that the phonograph would be primarily a personal *recording* medium rather than the mass playback device it became. And in the other, he failed to predict that people would acclimate themselves so thoroughly to the telephone's immediacy—its *live*-ness—that recorded conversations would eventually end up as rare exceptions to the rule, the province of journalists, wiretappers, and presidential historians.

Edison should probably be forgiven for these lapses. After all, "playback" and "live" media were cutting-edge concepts in the 1880s, so much so that the terms themselves hadn't been coined. (Before the telephone, experiencing "live" media entailed learning the obscure, syncopated grammar of Morse code, while audio "playback" was for the most part limited to the punch-card arrangements of the player piano.) Edison's bumbling with the phonograph reminds us, first, that the history of technological innovation is a history of happy accidents and brilliant mistakes. But it also suggests the perceptual difficulties that come from being too *close* to emergent technology. As they evolve, machines tend to throw off shadow versions of themselves, red herrings, false leads. (Radio began as a distributed, many-to-many, bottom-up medium, much like the early days of the World Wide Web, but it quickly consolidated into the broadcast model, dominated by national networks like RCA and NBC.) These false leads become even more baffling once you drop complex technologies into the whirlwind marketplace of postwar consumerism, where the couch potatoes and gadget freaks invariably stumble across new applications on their own. As William Gibson writes in *Neuromancer,* "The Street finds its own uses for things—uses the manufacturers never imagined."

Computers, as it turns out, are particularly vulnerable to these errant interpretations. Digital machines are born shape-shifters. Being digital means being able to reinvent yourself at the click of a mouse: morphing effortlessly from calculator to spreadsheet to word processor to videoediting console to battlefield and back again. It should come as no surprise, then, that the thinking machine's evolutionary path is littered with wrong turns and deviations. Some of these are errata of *scale,* such as IBM president Thomas Watson, Jr.'s legendary announcement in 1943 that the ultimate worldwide market for computers would be around five machines. Some are errata of *purpose,* where the technology's eventual application is misunderstood, as in the story about Edison's phonograph.

In the mid-seventies, several years before the Apple II first exploded onto the marketplace, an Intel engineer called a meeting of the company's board of directors to make an impassioned case for building a personal computer. He rolled out his vision of a future where consumers bought digital machines for their homes the way they currently bought televisions, stereos, and vacuum cleaners. The fact that Intel already possessed the technology—the chips, the integrated circuitry, the power supply—to make a machine for less than ten thousand dollars made the case a particularly compelling one, even though the behemoth mainframes of the day regularly sold for hundreds of thousands of dollars. But the board wanted an answer to a question that seems self-evident to us today: what were people going to *do* with these personal computers? Amazingly enough, the engineer didn't have a satisfactory answer: his most compelling scenario involved filing electronic versions of cooking recipes. Of all the eventual high-tech applications devised for the personal computer, all those spreadsheets and word processors and video games, the best he could come up with was a digital version of Mom's tuna casserole. It was like inventing the wheel and then immediately demonstrating what a wonderful doorstop it made.

The interface, too, has encountered its fair share of misappraisals over the years. We've already seen how the original Mac look-and-feel struck many commentators as being toylike, more suited to an arcade game than a Serious Business Application. It isn't that corporate successes like Windows 95 are *less* playful or cartoonish than the first Mac design (think of Microsoft's blue-sky desktop, or the animated characters of Office 97). If anything, the rise of high-resolution color monitors and faster screen-redraw technology has made the modern interface sprightlier, *more* toylike than its predecessors. What has changed is our expectations about what computer graphics are good for. In the mid-eighties, a fancy visual display could serve up a pie chart when necessary, but the dancing pixels were usually reserved for the Loderunner and Asteroids sessions. The graphic interface revolution has changed all that: we now intuitively understand that visual metaphors—all those blinking icons and desktop patterns and pull-down menus—have an important, and increasingly indispensable, *cognitive* function. They help us imagine our information, envision it all in one comprehensible vista, in a well-ordered landscape of data scrolling across our screens. No one mistakes computer graphics for mindless thumb-candy any more, for good reason.

For all its success, though, the graphic interface is still being misunderstood. Unlike the first generation of naysaying, today's misconceptions stem from *too much* faithfulness to the GUI's basic principles. There is a foolish consistency running through much of contemporary interface design, a blind spot at the center of Silicon Valley's usually acute field of vision. In a world dominated by icons and visual metaphors, the role of *text*—letters and words, rather than images and animations—has come to seem like an afterthought, an obscure walk-on part in a grand Hollywood epic. Words, in this lopsided paradigm, are always inferior to images. Anyone who knows anything about the history of writing systems—specifically the shift from hieroglyphic-style pictograms to phonetic spelling—will sense something bizarre in this hierarchy. Fortunately, the skewed priorities of the day are too imbalanced to last long. The *textual* revolution may well be the Great Leap Forward of interface design circa 2000.

The aversion to text has some understandable roots. In the language of New Age psychology, the contemporary interface is still in recovery, working through the post-trauma syndrome that has plagued it since it was first liberated from the command line. Computers began as number crunchers, but they spent most of their adolescence under the tyranny of text—all those inscrutable commands and instruction sets emblazoned on green-phosphor monitors and etched into punch cards. In its original usage, in fact, "interface" was just another word for text: input entered with keystrokes, output delivered studiously to the printer or the monitor. All the great languages that governed the relationship between computer and user were text-driven: BASIC, COBOL, Unix, DOS. Compared with the bitmapped universe of the ALTO or Windows 95, these older, textual experiences now seem lifeless and unintuitive to us—like a Technicolor movie replaced by a page-bound script. Why do anything with words on the screen, when words were the source of so much trouble in the old days? Good interfaces do away with text, the way good psychotherapists get rid of repressed memories and emotional blocks. Text commands were the great inadequacy of the computer's early years, its Achilles' heel—which is why it's only logical that the modern interface has such an adverse reaction to words on the screen. Call it the DOS complex.

But for all the real advancements of the GUI, a new generation of text-based interface tools promises to transform the experience of using a computer—assuming the high-tech world recovers from its case of logophobia. Rest assured, these new interfaces will have nothing in common with the clunky, obscure syntax of Unix and DOS. The C-prompt, thankfully, remains a dead letter. Instead of forcing the user to memorize arcane commands, like a schoolboy dutifully conjugating his Latin, the new textual interfaces will help users navigate through information, the way our desktop metaphors do today. The new textual tools will work in the service of what David Gelernter calls "topsight," like the crow's-nest vistas of the Victorian city. The command-line regime chained its users to the minutiae of edit modes and proper syntax. The new tools will be a way of seeing the whole.

Let's be clear about a few things, though. Even the most conventional modern interface uses text in many different ways. Words appear in the *content* of documents, of course—in word processors and presentation software and page-layout programs. They also continue to have an important role as *commands,* the tools we use to get the computer to do something. Pull down your run-of-the-mill menu and you'll still see textual descriptions of the available commands. (Floating palettes and command bars, however, usually rely on icons for this.) Words remain an important part of the interface's meta-information, the data we lean on to explore the infoscape of our hard drives. Imagine navigating through the Mac Finder or the Windows File Manager without the luxury of named files— just a soup of cryptic icons and nothing else. It's not for nothing that icons come with a text appellation tethered to their base, like a name tag stitched to a second-grader's sweater. Search requests—"find all documents with *Kissinger* in their title"—are an essential navigational tool, and would be utterly useless without text. (As we'll see in the next few pages, the new textual enhancements are actually derivations of the search commands found in most modern interfaces.)

So text already wends its way extensively through the modern graphic interface. Unfortunately, there's nothing innovative about the way these interfaces employ words: as the visual language has grown increasingly elaborate, with 3-D file overviews and Bob-style animated characters, the old-fashioned textual language hasn't changed much in twenty years. (The one exception is the rise of hypertext, as we saw in chapter 4.) From an evolutionary perspective, textual interface elements appear to have stabilized around a unified genetic makeup. But new adaptations are on the horizon. As it turns out, the most compelling evidence of this new species comes from unlikely habitats: the worlds of presidential politics and Elizabethan literary criticism.

In March of 1996, *New York* magazine published a controversial cover story claiming to reveal the author behind *Primary Colors,* the anonymously penned roman à clef about the early months of Bill Clinton's first campaign for the presidency. The book had been released just two months before, and speculation about the author's identity was still bubbling through the literary-political establishment—all of which meant that the editors of *New York* knew their claim would have to stand up under strict scrutiny from other members of the press. Still, they stuck by their story: *Newsweek* columnist—and former *New York* reporter—Joe Klein was almost certainly the author of *Primary Colors.* The accusation itself was not terribly surprising: Klein had appeared on a number of lists of possible suspects, and his tortured, love-hate relationship with Clinton matched the psychological profile of the author. What *was* surprising was the source for the story: not a leaked memo from the publisher, or a private detective rummaging through Klein's trash. *New York* had staked its reputation on a computer analysis of *Primary Colors,* implemented by a relatively obscure English professor at Vassar, Don Foster. After all the idle hypotheses and armchair sleuthing, the first authoritative pronouncement on the mystery had come from a machine.

How had a computer arrived at such a judgment? And why would anyone bother to pay attention to it? Digital technology has advanced mightily in the past fifty years, but it certainly hasn't advanced far enough to read a book and comprehend it, much less speculate on the author's identity. What business did a few insentient lines of binary code have participating in the great literary whodunit of the nineties?

The short answer is that computers don't need to *understand* a document to do useful things with text. (Your everyday spell checker is a case in point.) Millions of dollars and person-hours have been expended chasing the holy grail of AI: a computer that comprehends language, that follows semantics as readily as it does statistics. The world eagerly awaits a *meaning* cruncher, though all the evidence suggests that it will have to wait another decade or two for such a creature to appear. Fortunately, there's a shortcut of sorts, a hack. Language, it turns out, is not exclusively a semantic affair. There are also *statistical* properties to written language, properties that are beyond the perceptual skills of most humans (excepting the occasional idiot savant with a photographic memory).

Any text can be reduced to an inventory of words, arranged not by syntactical order but by frequency, like the concordances of biblical scholarship. Some of these numbers are more relevant than others: this chapter contains 687 instances of the word *the* and 11 of the word *interface*. Any English speaker will immediately comprehend that the latter number reveals more than the former. You might hardwire a computer to recognize the distinction by instructing it to disregard all articles, pronouns, and prepositions in its word inventory. But a better solution might be to allow the computer to reach the same judgment on its own by performing a comparative study of several documents—let's say, for the sake of argument, an excerpt from this chapter, a chunk of Apple's *Human Interface Guidelines* manual, and a selection from Stephen King's *The Shining*.

Plug all three documents into the computer and generate a word inventory for each. Ask the computer to eliminate the high-frequency words that appear in all three texts. If you have a large enough sample, you'll immediately toss out the bare necessities: the pronouns, prepositions, and articles littered throughout each document. What remains are the more distinctive words: *interface, bitmap, mouse, Overlook, blood.* In fact, the word inventory for *The Shining* excerpt is so idiosyncratic that it immediately stands apart from the other two. The computer—trained only to count words—can sense the difference between a Stephen King novel and this chapter, even if it can't begin to explain what that difference *means.* But the gap between this chapter and the Apple manual is trickier: the two documents share a number of high-frequency words that happen to be relatively unusual ones (*interface, click,* and so on). In the language of systems theory, these words are information-rich. Their rarity makes them meaningful, in the same way that the description "he has a nose" has less information than "he has a Roman nose." (Nearly everyone has a nose; not everyone's is Roman.) Nearly every document includes the word *the;* few contain *interface.* The use of *interface* tells us more about its container document, though to *see* that difference you need a broader sample of English-language use. Without that broader scope, the results tend to overemphasize the coincidences, placing undue importance on the terms that happen to overlap. (According to our limited sample, *bitmap* is a common word, since it appears in two-thirds of the documents.)

Let's be precise about the terminology here. When we talk about the computer "seeing" the differences between documents, that is not to imply that the computer actually *understands* those differences in a literal sense. Perhaps a better way of putting it is that the computer "registers" the distinction between *The Shining* and the Apple manual, because it can describe that distinction numerically. The numbers reveal something about the meaning of each document, even if that revelation is an oblique one, like a low-frequency tone shimmering in a puddle. The computer can't hear the meaning directly, but it can catch glimpses of it in its statistical reflections.

Despite their indirect origins, these glimpses can be remarkably informative, given the intense document-processing skills of the modern PC. We're accustomed to the computer's number-crunching skills delivering miraculously precise word counts and revenue projections, but what if that same power could be harnessed to pursue a more elusive quarry? Sure, the computer can sense the difference between a computer manual and a paperback fright-fest, but your average second-grader could probably do the same. (And all the computer can do is point to the *contrast* between the works; the second-grader would probably be able to tell you much more about their respective *meanings*.) But couldn't you redirect that numerical prowess toward a loftier goal? Is it possible to perceive higher-level linguistic attributes—meaning, style, intonation—through the lens of statistics? Can a machine *make sense* of language without learning how to read?

To answer that question, we need to venture back to one of the enduring mysteries of Shakespearean scholarship, a mystery that was solved by the statistical analysis of digital technology. Of all the arcane puzzlements of Elizabethan literary criticism, few have been as tantalizing, and elusive, as the details of Shakespeare's acting career. For generations, scholars have known conclusively that the Bard performed in every play that he wrote; in two plays, in fact, we know which parts he played: the ghost in *Hamlet* and Adam in *As You Like It*. His other roles, however, remain a mystery. Because records of the Globe Theater's production schedule have survived through the ages, we know the run dates of each play in the oeuvre. We also have a reasonably accurate chronology for his writing career, which means that we can gauge with some precision the overlap between performance and composition. In other words, we know that Shakespeare was writing *King Lear* while he was acting in *Othello,* and that he was acting in *The Merchant of Venice* while he was writing *Henry IV.* We just don't know the parts he was playing at the time.

Or at least we *didn't* know—until Don Foster stumbled across a brilliant, and strangely reassuring, idea. If Shakespeare had indeed memorized the lines for a part in one play while composing the script for another, then perhaps there had been a little seepage between the two. Perhaps the ritual of performing every night had lodged certain words in Shakespeare's head, like the detergent jingle from morning TV that hounds you through the workday. Anyone who writes for a living will recognize this phenomenon immediately. Words cycle through our daily vocabulary at different rhythms. Certain words stick with us for life, and remain immediately accessible to us at any moment: the names of loved ones, the building-block grammar of our native tongue, the primary colors and cardinal numbers, and so on. Other words wax and wane, in sync with forces larger than the individual speaking them: the fashionable vagaries of slang, the geek-speak of technological innovation, the "ethnic" idiom derived from broader demographic trends. (Think of the influence of black English on the mainstream American dialect over the past twenty years.) Most words, however, lie somewhere in between: drifting in and out of our regular vocabulary, like a band of itinerants cursed with a hankering to settle down. The word *profound* strays into your head and sits there for weeks, at the very edge of consciousness, primed for use. And for weeks, whenever a situation arises that demands a tone of seriousness or intensity or ironic overstatement—the word *profound* rolls out like clockwork. But soon enough another contender implants itself (*major,* let's say, or *crucial*), and *profound* retreats to the darkened wings of occasional use.

Foster's breakthrough was to assume that even the great Shakespeare might be prone to the same linguistic habits. Was it possible, Foster asked, that words from Shakespeare's memorized lines were accentuated in the Bard's vocabulary during the run of each play? Could the language of Shakespeare's acting career have infected his playwriting? It took a computer to answer the question, a computer specially programmed to track Shakespeare's use of statistically meaningful words—words that he used fewer than ten times in his career. The computer analyzed the distribution of these words on two levels: first, their appearances in individual parts (*Hamlet*'s ghost, say, or *Midsummer*'s Lysander); and second, their appearances in entire plays. If Shakespeare the actor was influencing Shakespeare the playwright, then certain plays would be littered with the vocabulary from a part in another, earlier play. The scattering of high-information words would be so light and varied as to be unnoticeable to humans, but the computer's prodigious pattern-recognition skills would track it down in a matter of hours—assuming, that is, that Foster's hunch was correct.

The results that came back from the lab turned out to be as precise and clearly defined as a fingerprint. Each play possessed a mirror-role in another play, revealed by the shared idiom of high-information words, like a family of orphans reunited by the science of DNA testing. In each instance the overlap followed the chronology of performance and composition. The results actually exceeded Foster's expectations. The analysis, as Edward Dolnick reported in *The Atlantic,* could be confirmed from a number of different angles: "It never assigns to Shakespeare a role we know another actor took. The roles it does label as Shakespeare's all seem plausible—male characters rather than women or children. The test never runs in the wrong direction, with the unusual words scattered randomly in an early play and clustered in one role in a later play. On those occasions when Foster's test indicates that Shakespeare played *two* roles in a given play—Gaunt and a gardener in *Richard II,* for example—the characters are never onstage together."

Using only the limited tools of word counting, the computer had solved a mystery that had eluded sentient, English-speaking scholars for centuries. The sterile number-crunching powers of the PC could now tackle more rarefied, nuanced problems, problems that had as much to do with the *meaning* of language as with its statistical base. Once again, we see evidence that technology rarely advances along a steady curve; instead, progress happens in a nonlinear, staggered fashion, with steady, incremental growth punctuated by sudden leaps forward. Take a bowl of water and gradually lower the ambient temperature in the room; for a stretch of time, the change is linear: the water gets colder as the temperature drops. But at a certain point a threshold is crossed—in this case the threshold of zero degrees Celsius—and suddenly you have not colder water but ice, a new property, fundamentally different from the preceding one. A slower machine, equipped to handle less textual information, is nothing more than a literary bean counter—good for generating a concordance for a single document, but not terribly sophisticated otherwise. But ramp up the processing power significantly—far enough to do a *comparative* study of word use in hundreds of documents, not just one—and you hit a threshold point, a singularity. The number cruncher becomes a literary sleuth, outsmarting tenured professors and armchair Shakespeare buffs. In his *Atlantic* article, Dolnick speculates that Foster's software is a "sign of things to come" for literary studies. But the promise of literary computing extends well beyond the obscure details of Elizabethan drama. By the end of the decade, most personal computers will sport a version of Foster's program as a basic tool in the human interface, as essential to the user experience as windows and icons are today.

Perhaps the most startling thing about the Foster study is the *simplicity* of the program he used. The statistical properties of language, after all, are not limited to word frequency. There are, in fact, hundreds of attributes that the computer can use to build a numerical model of a given text. Which properties you decide to track depends on what you're looking for. Let's say you're trying to gauge the relative complexity of a document.

You could have the computer monitor the length of each sentence; you could measure syntactical intricacy by tracking the number of clauses separated off by commas, em dashes, colons, and semicolons. Simply calculating the average letters-per-word would probably be enough to differentiate, say, *The Cat in the Hat* from *Minima Moralia*. A combination of all three might be sufficient to generate a useful complexity ranking for text documents.

If you're after something closer to authorial style, the complexity test would be only one weapon in a broader arsenal. Word-frequency studies, along the lines of Foster's Shakespeare analysis, can reveal a great deal about an author's "voiceprint," but by stacking up a document's vocabulary into a digital concordance, they ignore an essential property of language. Words, as we all know, come in sequences—and that sequential order is as much a part of an author's voice as the individual words themselves. Authors often lean on predictable clusters of words in their writing. Usually these take the form of clichés: we have coddled tyrants, centers that don't hold, rained-on parades, and so on. (In the computer world, for some reason, "robust" has become the modifier of choice for state-of-the-art operating systems.) Clichés, though, are low on information, since everyone uses them. But most writers have more idiosyncratic word clusters—a fondness for the phrase *insanely great,* perhaps, or a tendency to drop Michel Foucault's name every time the word *prison* appears in a sentence. These linguistic patterns can tell us a great deal about an author's stylistic signature.

As it turns out, Don Foster used a combination of these techniques to "solve" the *Primary Colors* mystery. The overall method was similar to the Shakespeare case, a kind of high-tech compare-and-contrast exercise. Where the earlier study compared the language of each play to the language of each part, the later study measured the text of *Primary Colors* against excerpts from the work of twenty top suspects: Joe Klein, *The New Yorker*'s Sidney Blumenthal, novelist Lisa Grunwald, and so on. Foster had the computer look not only at the overlap of low-frequency words, but also at word clusters. Most of the suspects would have used the words *perceptive* and *Clinton* with some frequency; but if a writer often used *perceptive* in the same sentence as *Clinton,* the computer would pick that up. The sequence of words was as important to the study as the words themselves.

As with the Shakespeare study, the results that eventually rolled in were indisputable. Joe Klein was the author of *Primary Colors.* The overlap between his published work and the anonymous novel was several times greater that of any other suspect. To his later chagrin, Klein emphatically denied the charges, swearing on his "reputation as a journalist" that he was not the author. As for the technical evidence, Klein dismissed it with a passing remark: "It was clear to me from the beginning of the book that the author, whoever he or she is, was very familiar with my work." He had failed to see how a *computer* could also become familiar with an author's works—familiar enough to identify other unsigned texts by the author with a precision that human readers sadly lacked. When, several months later, Klein held a press conference to fess up to writing the book, he dropped another subtle dig at Foster's computer analysis. He had waited, he explained, until "credible evidence" had appeared naming him as the author of *Primary Colors.* The computer study, apparently, hadn't made the grade.

Foster's detective work should make it clear that pattern-recognition software can do some extraordinary things with text documents, things that may even appear superior to human reading aptitudes. After all, computers only learned to draw pictures a few decades ago; that they are now doing advanced Shakespeare scholarship suggests just how far the technology has come. But Shakespeare and *Primary Colors* are just the beginning. In both examples, the computer was looking for *stylistic* resemblances—dusting, in a sense, for voiceprints. The *meaning* behind each text was for the most part peripheral to the experiment, hovering somewhere beyond the scope of the inquiry. Divining the authorship of anonymous texts makes for a wonderful trick, of course, but it's hardly an everyday task for most computer users. The great bulk of information-management problems revolves around semantic issues, not stylistic ones.

Fortunately, the pattern-recognition tools are not hardwired to literary sleuthing. They can be just as adept at perceiving semantic relationships between documents. It's not difficult to understand how this might work. Simply rigging up a list of the four most common "high-information" words would tell you a great deal about a given document's meaning. Consider the list for this chapter: *interface, textual, pattern, Foster.* Even in that abbreviated form, you can catch a glimpse of the chapter's subject matter. Powered by advanced pattern-recognition software and "robust" indexing tools, the operating systems of the future will roll out these keyword summaries on the fly, the way they now display a document's name, creation date, and file size. In early 1996, Apple began showing a functional demo of its new Finder software, in which all file directories include a category for "most representative words." As you change the contents of the document, the list of high-information words adjusts to reflect the new language. At first glance, this may seem like a superficial advance—a nifty feature but certainly nothing to write home about. And yet it contains the seeds of a significant interface innovation.

Up to now, file-management software—the metaform of the computer desktop—has dealt only with the *external* properties of documents: the number of bytes used, the last-modified date, the creator application. The user supplies the only datapoint in that mix that has anything to do with the content of the document: the file name. And even that convention has its limits. (How many times have you scanned through a list of inscrutable file names, trying to remember what you called that long-forgotten memo?) Apple's list of high-information words raises the stakes dramatically: for the first time, the computer surveys the content, the *meaning* of the documents on its hard drive—even if that meaning is condensed down to an abbreviated four-word list. Up to now, our file directories have been limited to the shells of our documents, the exoskeleton of storage information and user-specified names. Apple's new Finder was the first to peer beyond that outer surface, to the kernel of meaning that lies within. And it was only the beginning.

The idea of high-information words will be familiar to anyone who has used a search engine on the Web or an advanced "find" command like the one included in Microsoft Word. The difference, of course, is that in a keyword search, the *user* determines what the high-information words might be, instead of the computer. Often this process involves tinkering with the combination of words, so as to weed out extraneous information—or "noise"—from the search results. It's the digital equivalent of jostling the TV antenna to get better reception. A search on "Shakespeare" is particularly noisy—you get millions of documents back, far too many to do anything useful with. But run a search on "Shakespeare" and "Don Foster" and the image suddenly comes into focus. Instead of millions of files, you get ten or twenty. A few may turn out to be impostors (the transcript of a speech by an Oxford don named William Foster that happens to quote *Romeo and Juliet*), but most of the files will have something to do with the Vassar professor and his Elizabethan studies. Most Web surfers churn through these refinements several times a week, if not more. They know exactly what they're looking for—they just don't know what combination of keywords best represents the information they seek, which is why most search requests involve several iterations of trial and error.

The new text-driven interfaces turn this procedure on its head. For one thing, the computer itself can now discern the telltale characteristics of a given document and produce a functional list of keywords on its own. But there's more to it than that. As we saw in the *Primary Colors* example, these pattern-recognition technologies are fiendishly clever at *comparative* studies. In fact, their capacity for perceiving relationships among documents is primarily responsible for their eerie, almost-human language skills. Think about the way Foster hunted down his prey for *New York* magazine. He didn't have to laboriously describe the authorial voice of *Primary Colors* to the computer, or come up with a list of keywords that best represented the author's style. Instead, he had the computer figure out *on its own* what linguistic patterns best represented the book as a whole, by comparing them to a broad sample of documents, penned by the major suspects in the case. In fact, without that comparative backdrop, the relationship between Klein's public

•

i
n
t
e
r
f
a
c
e

c
u
l
t
u
r
e

work and his anonymous writings wouldn't be visible. The computer can't see the relationship without the control study of other, unrelated texts to measure against. Think about it this way: if you're trying to explain the connection between house cats and lions to an alien unfamiliar with the earth's animal kingdom, it's useful to have a few dogs and horses in the picture as well. Otherwise, the comparison lacks real resolution—the two cats look kind of like each other, and kind of not. The same goes for the computer's pattern-recognition skills. As you feed the machine more and more comparative documents, its "ear" for language scales up proportionately.

How will these pattern-matching technologies affect our day-to-day computer use? Consider what happens to the "search" or "find file" command alone. Rather than fumble around with the trial and error of a keyword search, you simply select a document and tell the computer to find some other documents that are similar to it. You decide what general criteria you want to use: you might want documents that cover the same topic but at a higher level of complexity, or a broad sweep of any documents vaguely related to the sample document, or an essay written with a similar sense of style. The software then configures its pattern-matching algorithms to fit the criteria you've selected, scans all the words at its disposal, and builds a profile of the reference text and its filiations with the rest of the database. The user says "find me anything like this document" and in return the computer dishes out a list of ten related texts—ranked by how closely each document matches the original—and also includes the "overlap" high-information words for each item on the list. Some resemblances may be statistically impressive but irrelevant to the user's needs. Let's say you're captivated by this idea of textual interfaces and search the Web for documents related to this chapter. You might get back a document about Don Foster's more traditional, low-tech Shakespearean criticism. (The two texts happen to have a lot of words in common, but they're the wrong words.) The feedback from the computer, however, shows you not just the general accuracy of the match but also the keywords that the two documents share. You can see at a glance that a document farther down the list—with *textual* and *interface* as overlap words—is a much more intriguing prospect, even though its overall score is lower than the Foster essay. In effect, the dirty work of refining the keyword search is subcontracted out to the computer. The user simply says "find me something like this" and the software does the rest.

So far, what we have here are a few rarefied academic successes and one dynamite search engine, but nothing that quite adds up to a paradigm shift in interface design. To understand the significance of these new tools, let's turn back to Apple's prototype version of the Finder. As it happens, Apple has already implemented all the above features in working, relatively stable code—though whether any of them will ever come to market is another story altogether. (And a sad one at that.) In the prototype Finder, Apple's pattern-matching engine—called V-Twin—goes far beyond your run-of-the-mill "find" command, becoming an essential part of the Mac's user interface, integrated into the ground-level information filtering of the desktop. It is here that the real revolution of text-driven interfaces should become apparent.

Apple's V-Twin implementation lets you define the results of a search as a permanent element of the Mac desktop—as durable and accessible as your disk icons and the subfolders stacked beneath them. In Apple's language, these new items are called "Views," for reasons we will come to understand. At first glance, a View looks and behaves like your average folder or subdirectory: it is represented by an icon; clicking on that icon opens a window that contains other icons, representing assorted files; clicking on one of those icons opens the appropriate document. So far, so good. Things get tricky, however, when we try to *add* a file to a View manually, by dragging an icon over the View's window. This probably constitutes the second most familiar mouse gesture in the human interface idiom (just shy of double-clicking), and yet when you try it on a View window, it doesn't work. The file zips back to its original location, rebuffed. At this point, you say: "Huh? I thought this was going to be an example of stellar, pathbreaking interface design. Wasn't the user experience supposed to be all about consistency? What's this window doing rejecting my icon?"

In this case, though, the old adage probably applies: you can't make an omelette without breaking some eggs. The fact that the View window departs so dramatically from the Mac conventions indicates how radical the shift really is, even if it seems innocuous at first glance. You can't add a file to a View window directly because the View is itself the result of a pattern-matching request. The user has only indirect control over the contents of a View. He or she specifies its general attributes, using the language of the V-Twin search engine: "find all documents on my hard drive that are *like* this other document." The computer then decides which documents fulfill that request and places them in the View window. (Technically speaking, it places copies—or "aliases"—of each document in the View; the originals remain in their previous locations.) Unlike the temporary results of a "find file" request, the View window has what programmers call "persistence." Like an ordinary folder, the View remains on your desktop until you throw it away. During that lifespan, the V-Twin software regularly updates the View's contents whenever new files arrive that match the original search request.

We've spent enough time wading through academia for our case studies, so let's roll out a corporate example to illustrate this technology at work. Let's say you're a lawyer working on an antitrust case for a telephone company. On your hard drive you have literally thousands of documents that might be relevant to this case: briefs you've filed, past decisions you've downloaded off the Web, interoffice memos, your own offhand notes. In addition to that body of work, you have thousands of other files that are clearly unrelated to the project at hand: briefs you've filed for other cases, past decisions that have nothing to do with antitrust law, memos circulated to you for no apparent reason, notes to yourself about grocery shopping or that novel you've been secretly planning to write. All those documents inhabit the labyrinthine world of your hard drive, and chances are you've organized them in a way that makes sense taxonomically but that directly obstructs your work on this particular case. In other words, you've got all the memos deposited safely in the "Memos" folder, and all the briefs cohabiting in the "Briefs" directory. But what you really want to see is all the documents that have anything to do with the antitrust case you're working on, no matter what generic category they belong to.

That's where the View window comes into play. Instead of burrowing manually through your file directories looking for pertinent documents, you select one file that represents the broad issues of the antitrust case and then define a

View window based on its textual characteristics. The computer ascertains the "meaning" of the reference document and dutifully scans your files for potential matches. Any document that fits the bill is deposited in the "antitrust" View window. That window remains on your desktop for months, and your computer continues to tinker with its contents, reflecting any changes that you make to your files. The first sweep through your hard drive picks up a chapter from your novel that includes an extended rumination on the great trustbusters of the early twentieth century. When you delete the passage weeks later, the file disappears from the View window. By the same token, the new memo you draft explaining the case's legal precedent immediately pops to the top of the View's file list, months after you defined the window's original characteristics. The contents of the View window, in other words, are *dynamic;* they adapt themselves automatically to any changes you make to the pool of data on your hard drive. That's one reason that the user can't directly manipulate the contents of a View: the computer controls the ebb and flow of files in the View window, shuffling documents back and forth as they evolve. The term *View* itself was supposed to reflect the transience and adaptability of this system. The word *folder* suggests a permanent, physical home for a document, while *view* is just a way of seeing—short-lived, ephemeral, a passing take on the world that quickly gives way to another.

But the real shift here is not the idea of dynamic windows. If there is a genuine paradigm shift lurking somewhere in this mix—and I believe there is—it has to do with the idea of windows governed by semantics and not by space. Ever since Doug Engelbart's revolutionary demo back in 1968, graphic interfaces have relied on *spatial* logic as a fundamental organizational principle. A file appeared in a given folder because someone put it there. As we saw in the "Bitmapping" chapter, there was something genuinely liberating about this new spatial imagination—given the historical limitations of the command-line regime and the innate human aptitude for visual memory. But that revolution had its limits. With thousands of files on our hard drives, and countless more lurking just offscreen on the Web, organizing files by the old-fashioned spatial method doesn't make sense any more. The graphic interface depends upon this one assumption: it still remains useful to say *I put this file here, and here it shall remain.* We know from our initial discussion of digital technology that this spatial convention is a total fiction. The data behind a given document is in fact scattered aimlessly across the magnetic surface of a hard drive; that it is represented by a single, unified icon on a desktop is an artifice, a visual metaphor. It could just as easily be represented by twenty icons. For all intents and purposes, the data has no physical place in the world. Its spatial coordinates on the desktop are simply an illusion, a trick of the eye.

What the View window does, in effect, is say this: why not organize the desktop according to another illusion? Instead of space, why not organize around *meaning?* Perhaps the whole idea of a document existing in a physical location—in a folder, say, or on the desktop—is just a hollow legacy carried over needlessly from the limitations of the real world. Perhaps we simply required those spatial coordinates to ease us across the digital frontier, and now that we've acclimated to the new environment, we can shed those old habits of thought, like the space shuttle tossing off its solid rocket boosters after a successful launch.

We like to pretend that our file-management tools are driven by meaning: that's why we give our folders names like "Journal Entries" and "Client Information," but in reality each of those folders is defined by a simpler, and more tautological, axiom: *the files in this folder are here because I put them here*. That's all the computer knows about their relationship. This sounds perfectly intuitive, of course, because it's the way things work in the real world. But as we saw in the case of Bob and Magic Cap, there are some limitations of physical space that are best left behind—that's why our word processors don't reproduce the stuck keys of mechanical typewriters and our photo-retouching software doesn't simulate someone accidentally opening the door as you're developing a print. There are some "features" of the real world that are better lost in translation. Perhaps the idea of a document having a single location is one of them.

How would a semantic interface actually work? On the most basic level, you would still have some control over the contents of your folders, but that control would be more indirect, trickling down through the pattern-matching software of the interface itself. Some folders might be defined by simple attributes (any documents containing the word *pumpkin*). Some might be defined by comparative studies (any documents that are similar to the document "Pumpkin Pie Recipe"). You might also let the computer divide up all your documents according to the semantic relationships that it perceives in the mix. But whatever combination of organizational principles you followed, two things would make this system dramatically different from the way things work today. First, the whole idea of "one document, one location" would disappear altogether. Your pumpkin pie recipe might pop up in twenty different folders, depending on the general attributes you've chosen. All these different appearances would point back to the same text file, of course, so you wouldn't have the version-control problems that can come from making duplicates today. These multiplying locations might be bewildering at first, but it's not hard to imagine our getting used to them over time. In fact, the most bewildering part of the mix might be the necessary overlap between the older spatial models and the new semantic ones. After all, a document that is only defined semantically might simply disappear from view, if none of the pattern-matching requests were close enough to its linguistic characteristics. You'd have to have some sort of backup system, in case the specific word inventory of one document fell through the semantic netting, a system designed to keep at least one copy of each document accessible at all times.

But the most significant—and the most troubling—change would be this: a semantic file system would give the computer much more control over the organization of your data. You'd define the broad categories, but the machine would make the hard decisions of what goes where, including—inevitably—judgment calls that are better left to humans. The question of whether this newfound authority should be seen as an interface advance or a major step backward extends well beyond the field of text-driven interfaces. It is, in fact, at the very heart of the debate over intelligent agents currently raging in the high-tech world. The next chapter explores the parameters of this debate.

AGENTS

In the fall of 1816, a few months after the publication of his ambitious first novel, the German writer E. T. A. Hoffmann published a short story called "The Sand-Man" in a small literary periodical based in Berlin. In an age dominated by charismatic eccentrics, Hoffmann was a real curiosity: a failed composer and chronic alcoholic whose fiction veered between gothic hauntings and psychological obsession and displayed an eerie fascination with automatons. (Picture a hybrid of *Jane Eyre* and David Cronenberg condensed down to ten pages and you get the idea.) Of all Hoffmann's creations, though, "The Sand-Man" remains the most enduring, though its path to canonical status was fittingly oblique. The story itself involves an edgy, high-strung "unreliable narrator" named Nathaniel, a sinister carpenter with a fondness for poking people's eyes out, and a beautiful young princess—Olimpia—who may or may not be a mechanical doll.

"The Sand-Man" unsettles on a number of levels, and its serpentine route to critical success reflected that strangeness. In a rare excursion into literary criticism, Freud dashed off a minor essay about Hoffmann's story while working on his masterpiece from the postwar era, *Beyond*

173

the Pleasure Principle. The essay—"The Uncanny"—recast Nathaniel as a victim of Freud's newly discovered repetition compulsion, returning endlessly to the "primal trauma" of those lost eyes, the way veterans of the Great War compulsively reenacted the bloodshed and tumult of the Somme or the Ypres offensive. Like Hoffmann's original story, "The Uncanny" was elusive and idiosyncratic; it drifted erratically from personal anecdotes to wobbly literary criticism to hardcore psychoanalytic theory. It was obscure enough to be ignored by two generations of scholars—a footnote to Freudian history, a minor work—until the post-structuralists of the sixties and seventies rediscovered it, and through it rediscovered "The Sand-Man." According to the post-structuralists, Freud's essay was a veritable case study in semantic undecideability, wavering between a coherent analysis of Nathaniel's neuroses and a confession of guilt—psychoanalysis revealed to be yet another interpretative sham, like the dolls and automatons of Hoffmann's fiction. Having conquered the literary theorists, "The Sand-Man" slowly replicated from syllabus to syllabus, until the story became a legitimate classic in the short-fiction genre.

There's nothing terribly unusual about Hoffmann's roundabout path to the Great Books pantheon. (It took Dickens nearly a century, after all.) But both the psychoanalytic and the deconstructive readings have done "The Sand-Man" a disservice. This turns out to be one of those instances where an engaged familiarity with machines, and particularly digital machines, can help us make better sense of literature—contrary to the doomsaying of Sven Birkerts. Knowing something about the history of computers makes you a better reader of Hoffmann. Why? Because "The Sand-Man" is the first great literary expression of a theme that winds its way through twentieth-century narrative: the danger—and the allure—of mistaking machines for humans. "The Sand-Man" is one of the founding texts of a tradition that goes all the way to H. G. Wells, *2001*'s Hal, and the replicants of *Blade Runner.* By this late date in literary history, the automaton-passing-for-human trope is as familiar to our collective storytelling habits as the mistaken-identity tropes of the sitcom or the false endings of the whodunit. Every other sci-fi flick toys seductively with cyber-androgyny, and even highbrow fiction has dabbled in the blurring of human and machine, most notably in Richard Powers's wonderful novel *Galatea 2.2.* Once an obscure addendum to the Romantic tradition, Nathaniel's robotic obsessions—and the anxiety they unleashed—have become a fully fledged narrative cliché in the last years of the twentieth century.

But the anxiety surrounding human-machine hybrids is more than just a literary device. The eerie attraction of the mechanical doll in Hoffmann's story, the self-doubt of *Blade Runner*'s replicants—these are imaginative themes that also lie at the epicenter of contemporary interface design. Hoffmann wrote at the dawning of the industrial age, at a time when Europe seemed besieged by a new species of mechanical devices—more dynamic, more animated than anything before them. It was almost impossible not to find something demonic in that automated movement, and equally impossible not to be mesmerized by it, which is why so many of the first visitors to Manchester returned with a difficult mix of repugnance and admiration. Hoffmann's story is an attempt to translate that anxiety—the push and pull of the organic machine—into a creative medium. *Blade Runner* and *2001* attempt the same with film. These novels and movies and short stories are all attempts to make sense of the new hybrids of human and machine, to translate that volatile mix into a creative form that is somehow less threatening, that offers some catharsis or release to the audience.

Over the past ten years, interface designers have grown increasingly intrigued by these issues, to the extent that digital automatons have become a major topic of discussion within the industry. These designers are, in a sense, the imaginative heirs of Hoffmann, working through the risks and the promise of mistaking machines for humans, resolving our anxiety about such a mix through works of profound creativity. The legacy of Hoffmann, though, is only now becoming clear. For the first twenty years of interface design, the dominant model was architectural: interfaces imagined binary code as a space, something to be explored. The new interface paradigm brings us closer to Olimpia's glassy stare: instead of space, those zeros and ones are organized into something closer to an individual, with a temperament, a physical appearance, an aptitude for learning—the computer as personality, not space. We call these new creatures—these digital "personalities"—agents.

What is an agent? There are several definitions. The general concept dates back as far as the fifties (John McCarthy's Advice Taker software), but the term itself did not enter high-tech parlance until several decades after that. In 1989, Apple released a celebrated video entitled *The Knowledge Navigator,* which deposited a genteel, tuxedoed actor in the upper-right-hand corner of a PowerBook. Here began the stereotype of agent as digital valet: the video showed the end-user—a professor researching rain forest depletion—casually dispatching his pixelated assistant to find famine records in the campus online database, like William Powell ordering another martini. The butler metaphor had a certain aristocratic charm to it (the high-tech trades bubbled with Jeeves jokes for a few years), but the bow tie was only a sideshow, a distraction. The great legacy of *The Knowledge Navigator* lay in the fact that the info-butler was a person at all.

Apple's digital Jeeves inaugurated a tradition of agent-as-anthropomorph. At the time, the legacy made perfect sense; the Macintosh had ushered in the entire rhetoric of visual metaphors: the desktop, the trash can, the folder, the mouse. If you could transform electrons into inanimate objects, then why not aim a little higher on the evolutionary chain? Why not imagine the computer as a person—chatty, obliging, perceptive? Most of us talk to our machines anyway, cajoling them to download files, cursing them out after the floppy drive fails. Why not endow the computer with an equivalent response mechanism? If we're going to be talking to our PCs, we might as well give them the opportunity to talk back. And so the anthropomorphic agent was born: a little human trapped behind the glass of the monitor, bound to your every whim, like a trapped fly or a genie. In the eight years since Apple's video first appeared, the "personality-equipped" agent has seen disappointing results at the box office, despite the early critical acclaim. Both Microsoft and Apple tinkered with anthropomorphs in released products (the lumbering houseguests of Microsoft Bob; Apple's speech-recognition software), but the reception was for the most part indifferent.

Agents, however, don't always require personification; they can just as easily take the shape of a Web browser, or a dialog box, or a text document. Some the most promising agents toil almost entirely behind the scenes; like superheroes or multinational CEOs, their invisibility is the source of their power. Agents also differ in their preferred habitats. Some of them are shut-ins and sycophants: they settle into your computer's hard drive and stay there for good, watching your behavior and helping out when they get a chance. Other agents are full-time tourists, roaming across the Net in search of information and trudging back home only when there's news to report. Some agents are extroverts; they compile relevant data for you by chatting with other agents, swapping stories and recommendations. These three classes represent the range of possibilities for agent-driven interfaces: the "personal" agent, the "traveling" agent, and the "social" agent. Each implies a different understanding of human-computer interaction, for reasons we will come to understand.

Choosing one variation over the others has consequences. As agents infiltrate more and more of our daily lives, those consequences will extend far beyond the routine tasks of managing files or making plane reservations. Agents may turn out to have a profound effect on the way popular tastes come into being, much as the rise of the blockbuster movie changed our relationship to the cinema and the serial novel changed our habits of reading. As with much of the modern interface, design choices for agents will ripple out into the larger culture, transforming regions of experience that otherwise seem to be disconnected. Part of McLuhan's point in *Understanding Media* was that new technological forms—the television, the radio, the book—transform not only the balance of power between our senses but also our experience of other media. "Radio changed the form of the news story as much as it altered the film image in the talkies," McLuhan argued. "TV causes drastic changes in radio programming, and in the form of the film or documentary novel." Because agents are the most independent—the most autonomous—tools in the interface repertoire, their influence may turn out to be the most far-reaching, and the most subtle. That's one reason that the design of our intelligent agents shouldn't be left up to the CEOs and the technocrats.

One of the puzzling things about agent software is that the surface representation of the agent itself is so malleable. It is a general assumption of this book that our visual metaphors are as important as the underlying functions they signify. On the face of it, agents would seem to be an exception to this rule. The three types of agent—personal, traveling, and social—can be represented in many different ways: a butler, a talking dog, a text report, a "personalized newspaper." With intelligent-agent technology, the visual metaphor is not as important as the underlying behavior of the agent itself. (Does it really matter whether it's a dog or a robot who's watching the stock market for you?) But this raises another question: does that underlying behavior really belong to the realm of interface at all?

The answer to that question gets to the essence of intelligent agents. As we saw in chapter 1, the modern graphic interface is defined by "direct manipulation." The user makes things happen in an immediate, almost tactile way: instead of telling the computer to delete a file, you drag it into the trash can. The underlying event is the same (the CPU follows instructions to erase a few sectors of the hard drive), but the illusion of the graphic interface is that you seem to be doing the work yourself. Much of the GUI's celebrated ease of use derives from its aptitude for direct manipulation, to the extent that more roundabout ways of working with data are usually frowned upon. But agents don't play by those rules. They work instead under the more elusive regime of indirect manipulation.

As the name suggests, agents are delegates, representatives. They do things for you. Being able to delegate responsibility to a software agent can be enormously liberating, but it comes at a price. The tactile, unmediated control of the traditional interface gives way to a more oblique system, where your commands trickle down through your representatives, like a desktop bureaucracy. Take, as an example, one of the most rudimentary tasks of the "personal" agent: emptying the trash can (or the "recycling bin" in Windows 95). Let's say you're the type of computer user that likes to pile up digital garbage on your desktop before throwing it out. Every few hours you toss another item in the trash can, but you don't empty it until you've got several megabytes of data waiting to be incinerated. A personal agent would observe this behavior, and after a few iterations, it would politely intervene and say, "I've noticed that you tend to empty the trash after it reaches two megabytes. Would you like me to do this for you automatically?" The agent might also decide to play Felix to your Oscar and volunteer to toss out your junk more regularly. In either case, though, the agent performs a task that was once executed by direct manipulation, by a mouse gesture or a menu command.

Intelligent-agent partisans consider that shift to be an enormous advance in ease of use: the only thing easier than directly manipulating the trash can is letting someone else do it for you. It seems intuitive enough, but there are dangers in ceding that additional control to the computer. The original graphic-interface revolution was about empowering the user—making "the rest of us" smarter, and not our machines. Agents work against that trend by giving the CPU more authority to make decisions on our behalf. It's this new authority—and not the representations of cartoon puppets or digital butlers—that endows the intelligent agent with its "intelligence."

The question of whether this intelligence is a good thing turns out to be a very old and mysterious one. Interface designers are hardly the first to suggest an answer to it. A decade after Hoffmann published "The Sand-Man," another European Romantic published a book about the perils of endowing inanimate matter with intelligence. It was called *Frankenstein*.

If the debate over intelligent agents mirrors the plot of Mary Shelley's novella, then Jaron Lanier is rounding up the villagers and storming the good doctor's castle. Lanier—the dreadlocked musician and virtual reality inventor—has been waging a prolonged rhetorical battle against agents for the past two years. In early 1995 he published a short rant on the Web entitled "The Trouble with Agents," subsequently expanded into "Agents of Alienation." A year later, he took on MIT's Patti Maes (the godmother of the "social" agent) in a heated debate about agent technology featured in HotWired's Brain Tennis section. The subject is obviously one about which he feels a great passion. "The idea of intelligent agents is both wrong and evil," Lanier writes, with characteristic intensity. "I believe that this is an issue of real consequence to the near term future of culture and society." These are strong words to level at an animated puppy and a guy in a monkey suit. Why direct such hostility toward a software product? Aren't there more offensive cultural antagonists to thunder against?

Yes and no. Lanier may be exaggerating the malevolence of the intelligent agent, but he's not overstating the significance of the issue. Like so much of the recent high-tech debate, Lanier's polemic reduces a complex issue down to an easy, and misleading, opposition between "good" interface design and "evil" intelligent agents. As in the Sven Birkerts versus John Perry Barlow dispute that ran in *Harper's* in the summer of 1994, the polarities obscure as much as they reveal. Agents will turn out to be many things, and their effects will ripple through the infosphere in multiple ways. Some of those effects will be regrettable; some will be genuinely encouraging. New technologies rarely speak with a single voice—that's one reason that we're so captivated by them. What is certain, though, is that the effects of these new technologies will be far more profound than we imagine them to be now. And it is on this front that Lanier's jeremiads do us a great service. They push the popular debate toward the proper scale. Instead of arguing over the feature sets in the latest spreadsheet package, we're at least talking about the broader cultural effects of new technology. That alone is reason to take Lanier's objections to the intelligent agent seriously.

But to understand those objections, we need first to clarify which agent we're talking about. For the most part, the "personal agent" doesn't belong on Lanier's hit list. These agents are more Igor than Frankenstein. They skulk around the lab, lending a hand for only the most rudimentary tasks: emptying the trash can at regular intervals, sorting your e-mail, running virus software. The full-page ads in *Wired* will try to tell you that these programs are real "intelligent agents," but don't be fooled. Sure, they live by the rules of indirect manipulation, but their so-called intelligence is just a marketing campaign, a slogan. They subsist on a small trickle of delegated authority, which is both a curse and a blessing. An agent granted the right to back up your hard drive without asking you first is not going to stumble across a cure for cancer along the way—but by the same token, he's not going to peremptorily decide to erase your disk either. Reining in the intelligence of your personal agent makes sense, particularly at this early stage in the technology's development. You don't want your representatives tampering with your data without your express permission—even if that tampering might lead to interesting results.

No, the temptation comes from having your agent tamper with other people's information. The real threat lies in the "traveling" agent, the one that unmoors from the host computer and strikes out for the terra incognita of cyberspace. As Lanier puts it: "An agent's model of what you'll be interested in will be a cartoon model, and you will see a cartoon version of the world through the agent's eyes. . . . Advertising will transform into the art of controlling agents, through bribing [or] hacking." The traveling agent imagined here is probably the most clearly realized model of the agent-as-representative idea: the agent represents you in its dealings with other agents, shifting through an entire repertoire of personae over the course of a day: a clothes buyer one moment, a personal secretary the next. The first commercial version of this software arrived with much fanfare in late 1994, when a Silicon Valley start-up called General Magic introduced its Telescript programming environment. Cofounded by Bill Atkinson and Andy Hertzfeld (two of the key visionaries behind the original Mac), General Magic was widely hyped as the agent-driven software company of the future, and Telescript was going to be its DOS: an operating system for agents.

Telescript, technically speaking, was a communications protocol, not an operating system. It provided a common language that agents could rely on when negotiating their deals, a lingua franca for the 'bot community. On paper, it was an enormously seductive product. Imagine a scenario in which a user dispatches a software agent to a financial service to monitor the vicissitudes of a given stock, instructing the agent to sell after the stock reaches a certain price threshold. The agent attaches itself to the financial network, where it follows the stock for weeks or months without communicating back to the user. Once the stock reaches the preordained value, the agent executes the "sell" command, depositing the proceeds into the user's bank account. The agent then travels to an online travel agency, searching for the cheapest round-trip ticket to the Caribbean. The agent purchases the most economical fare and returns to the user's computer, entering the flight itinerary and checking the user's monthly planner for any potential conflicts. If a meeting is scheduled during the vacation, the agent promptly notifies the other participants via e-mail and arranges a new time amenable to all.

On a technical level, the magic behind Telescript relied on a fundamental change in the way network communications operate. The existing model went by the name of remote procedural calling (or RPC); Telescript proposed a new regime called remote programming (RC). In the RPC system, retrieving information from an online service requires a direct connection between client and server. Because the information content of the server is constantly changing, there's no way of determining what the latest data is without dialing up the service and browsing through the available files. It's an extremely inefficient system for both user and network. If you're waiting for a particular kernel of data to come down the wire (a press release, perhaps, or the results of a cricket match), you must log on repeatedly until the desired information is uploaded onto the network, wasting time and money in the process. The network itself is consequently forced to handle increased traffic, as countless users prowl about the server waiting for their data to appear.

The remote-programming model implemented by Telescript did away with these redundancies. Instead of logging on every fifteen minutes, the user communicates with the network in short but efficient sessions. If you're eager for the latest word on Bosnia, you dial up your Telescript-compliant online news service and quickly upload a Telescript agent that has been instructed to find any bulletins containing the keyword *Bosnia*. This agent is not an inert document like an e-mail message, but rather a miniprogram of sorts, one that will continue to run on a remote hard drive long after you've disconnected from the service and returned to more pressing matters. The agent monitors the news feed twenty-four hours a day while residing in its new host environment (called a "place" in the Telescript lingo). The moment a Bosnia-related story arrives, the agent dials up your computer, deposits the story on your hard drive, and announces a successful conclusion to its mission.

In this scenario, the entire exchange between client and server unfolds in two brief stretches of direct communication. From the user's perspective, the processing time between the initial instruction (in Telescript terms, the "go" command, dispatching the agent to the remote site) and the desired result (the agent returning with the story) takes place entirely behind the scenes. No time is wasted searching for files that have yet to appear. The online service may in turn charge a nominal fee for providing the agent with accommodations during its stay on the remote server. General Magic proposed that this duration be recorded by the agent in "teleclicks"—a new form of e-cash designed specifically for Telescript networks.

These hypothetical scenarios of Telescript at work sound, at first, like a case of wish fulfillment for info-junkies, like having a personal secretary and research assistant without the payroll expenses. But Telescript's model goes beyond the notion of agent as human surrogate. An online service populated by intelligent agents looks less like a massive storage device than it does an ecosystem teeming with digital life-forms, each pursuing a separate goal. A server bustling with self-sustaining agents resembles an ant colony more than an old-fashioned machine, a gene pool more than a cotton gin. It is partly this organic quality that alarms Lanier. He sees in these new agent colonies a sinister extension of Madison Avenue's reach: "Today's advertising agencies will become tomorrow's counteragent agencies. This might involve fancy hacking, but it might also be softer. Counteragencies will gain information about agent innards in order to attract them, like flowers wooing bees. Regular netizens won't have this information, so they will attract no bees and become invisible."

While Telescript was a brilliant work of programming, it barely survived its commercial launch—though the reasons behind its failure had nothing to do with Lanier's objections. Communications problems undermined Telescript—for a "traveling" agent, it was surprisingly illiterate in the languages of the wider world. Just as the Web was lurching into the popular consciousness, General Magic had bet the farm on a proprietary communications protocol, one that wasn't compatible with the open standards of the Internet. That alone was the deathblow for Telescript. (General Magic claims to be working on a Web-compatible version of Telescript, though its future is very much in doubt.) But autonomous agents like those envisioned by Telescript will soon appear on the Net, one way or another. Will they result in Jaron Lanier's brutal ecology of agents and counteragents? Or is there another, less menacing alternative?

In the long tradition of cyborg literature, the one that stretches from Hoffmann to Philip K. Dick, the hybrids of human and machine usually run into trouble when they get tangled up in human desire. Think of Nathaniel's obsession with the mechanical princess; the cinematic Frankenstein's longing for a bride; Harrison Ford's tortured fixation with the replicant women in *Blade Runner.* As long as our cyborgs restrict themselves to straightforward tasks, our relationship to them remains relatively untroubled. But once they infiltrate the murky, psychological realm of taste and desire, all hell breaks loose. The human can't help but project those appetites and longings onto the machine, and the machine invariably returns the favor, or at least appears to. (The process is not unlike the transference and countertransference that Freud described in psychoanalysis.) The lesson of most cyborg literature is that the commingling of human and machine in the life of the libido is inevitable, and inevitably destructive.

A comparable theme runs through Lanier's critique of the intelligent agent. As long as the agent only executes clearly defined commands, it's hard to see a problem. The more mathematical the commands, the better. An agent dispatched to buy a plane ticket for less than five hundred dollars or a technology stock with a price-earnings ratio below ten— this sort of agent hardly poses a threat to "the future of culture and society." Like the stay-at-home personal agent that only empties the trash can for you, traveling agents that follow specific, quantifiable rules should be seen as a worthwhile software advance, assuming the programmers can come up with a common language to unite them.

The ultimate goal of the more ambitious agent enthusiasts, however, goes well beyond software that dutifully does what it's told to do—book airline tickets, sell stock. The real breakthrough, we're told, will come when our agents start *anticipating* our needs—the intelligent agent that makes an appointment with the nutritionist after noticing the number of pizza delivery charges on the monthly credit card bill, or has flowers delivered a day before that anniversary you're always forgetting. As with so many of these vaporware products, the scenarios sound perfectly appealing in theory, but you can't help suspecting that the shipped product—if it ever arrives—will not work as smoothly as advertised.

As in the world of espionage, the central problem with intelligent agents is that it's not always clear who they're working for. The desktop-based info-butler—the one that follows your orders without improvising—is clearly working in your service, but as more and more of our computers are permanently connected to the Web, new types of agents will appear, agents that dwell on other servers and migrate out to your machine only when they sense that they might be useful to you. Let's say you're a devoted baseball fan and your online surfing history shows it; you're a regular visitor to fastball.com, where you review the latest box scores for the Chicago White Sox and post frequently in the discussion forum. An agent residing on the fastball.com server picks up on your fondness for the Chisox and assumes from these activities that you might be interested in breaking-news updates on the team: Albert Belle breaks an ankle, or Doug Drabek tosses a no-hitter, and the fastball.com agent dispatches an alert directly to your desktop.

This sort of intelligent agent also goes by the name of "push media." In the most basic sense, it is information that comes to you, as opposed to information that you go get yourself. (Traditional Web browsing is accordingly thought of as "pull media.") The "push" agent at fastball.com tracks you down to notify you about Belle's twisted ankle; you don't have to sit around all day reloading the "injury report" page on your Web browser, waiting for something to happen. The relevant information is pushed to you directly—which means, in theory, that your needs are satisfied before you even realize they exist. Everything that arrives on your desktop will be tailored to your "personal tastes and preferences," a custom-fit delivery of information and services that's always one step ahead of you.

As enticing as it sounds, the trouble with that baseball agent is that its allegiances to you are only provisional; its real "master" is the fastball.com site. And this is why push-style agents can be a frightening thing indeed—although not frightening in the usual sense of *X-Files* paranoia, where all your actions are monitored by the shadow state of "push." (At least the computer virus panic had a certain technobiological cunning to it.) The threat of push is much more mundane. It's the threat of personalized schlock, detritus, data-smog posing as your new best friend. Don't think information butler. Think junk mail.

I don't know about the folks in Silicon Valley, but the direct marketing that slides under my door every morning seems to have a pretty good sense of my "personal tastes and preferences." You could build an entire personality profile just by shuffling through a week's worth of my mail. You can discern my political leanings, my profession, my tastes in clothes, my computer platform, my neighborhood. The push advocates talk rhapsodically about the merits of tailor-made information, but the carbon-based pushers of direct marketing seem to be keeping pretty good tabs on all of us already. Just because it's customized junk mail, it doesn't mean it's not junk. When I think about the future of push agents, what comes to mind immediately is that crowning achievement of late-twentieth-century democracy: the direct-mail solicitations from political candidates. They began as form-letter giveaways, with blocky, dot-matrix typefaces and corporate return addresses. But slowly the genre evolved: first there was faux handwriting from the candidate himself, then the typewritten address of a personal correspondence, then the solemn appeals handcrafted to push your demographically calibrated button. The increased precision translates directly into increased irritation. The only thing worse than receiving a piece of junk mail is being duped into opening one.

There will be similar forces at work in the new economy of intelligent agents. The pushers will strain mightily to transform their ads into "information services" by making purportedly educated guesses about our every passing whim. Consider this example from *Wired*'s euphoric February 1997 cover story on push media: "You are standing on a street corner of an unfamiliar city where you are attending a convention. On your PDA, you stare at a map of a city. It looks like rain. The weather icon starts blinking. Droplets pepper your glasses. On the map, tiny umbrella icons appear showing stores within a two-block radius that sell rain gear." It sounds reasonable enough, but the truth is, this plucky little PDA has made a formidable leap of logic with those umbrellas. By gliding effortlessly over it, *Wired* neatly sidesteps the real threat posed by agents that anticipate instead of obey.

This is the threat of letting your *computer* decide what you want, instead of telling it directly. *Wired*'s hypothetical machine senses rain and thinks rain gear, but it could just as easily think a thousand other things: book tickets for Caribbean vacation; cancel tee-time; order new aluminum siding; search local classifieds for SWF who "likes walking in the rain." You'll either get a random sampling each time it rains, or they'll all come tumbling down the wire at once. In either case, the signal-to-noise ratio will be like a sound check at a Sonic Youth show. Either you'll spend all your time sorting the junk from the not-quite-junk, or you'll just toss the PDA in the wastebasket and dance off, Gene Kelly–style, into the downpour.

As Lanier has argued tirelessly over the past few years, so-called intelligent software is usually an excuse for poor interface design. I don't really want my computer guessing what information I'm looking for—particularly if those guesses are being bankrolled by the marketing departments at Nike and Microsoft. What I want is a better way to get to that information. We need better road maps for information-space, not better home delivery. *Wired* sees an umbrella behind every raindrop. I see a Domino's guy ringing the doorbell every half hour, saying, "I've got a hunch you want a pizza."

And you know what? The Domino's guy might just be right. Who among us can say no to pizza? But I'd rather place the call myself, thanks. You need only look at AOL's problems with junk mail to see how quickly push comes to shove when advertising budgets are involved. What we really need are better ways to *pull*. That's what compelling interface design has always been about. It's the only way to keep the unwanted solicitations off our desktops. The pushers conjure up a future where every passing advertisement is mission-critical to your needs, where every billboard knows your name. But what if what you want is a world with fewer billboards?

Intelligent agents are not just in the business of anticipating our needs. They are also beginning to infiltrate the more nebulous realm of taste and aesthetic distinction. That we can even contemplate such a possibility should be a reminder of how far computer science has advanced in recent years. After all, deducing your interest in baseball news from your frequent visits to fastball.com is a simple enough logical leap. (You can imagine an elemental if-then scenario: if x visits more than y times, start sending x live updates on breaking news.) But an agent that could evaluate your tastes in movies or wine or even people, an agent that could build a nuanced model of your aesthetic or interpersonal sensibility—*that* would be a paradigm shift that lived up to the phrase. Only we know from the literary prehistory that there are always difficulties when desire enters the picture, when the agent starts coming up with ideas on its own, based on its appraisal of your personal tastes. Instead of buying that plane ticket to Palm Beach, the agent recommends a few secluded beaches in the Caribbean, given its knowledge of your recreational history. Rather than dutifully buy the tech stock you'd had your eye on, the agent consults its database of past acquisitions and comes back with a smattering of petrochemical issues. The trouble begins when our agents start meddling with our subjective appraisals of the world, when they start telling us what we like and don't like, like an astrologer or a focus group. This is where Lanier sees the dance of the bumblebee emerging, all those counteragents luring regular agents with their info-pollen.

But before that little ecosystem can come into existence, our agents have to learn something that genuinely resists simple classification: our subjective reactions to that outer world of music CDs, vacation getaways, people. An agent that can't convincingly demonstrate an understanding of our cultural appetites doesn't pose much of a threat, since we're not likely to entrust it with difficult decisions. (It can bargain-hunt for plane tickets all it wants, but we're not about to send it out shopping for a spouse.) For Lanier's dystopian future to come to pass—a Net populated by dumbed-down agents, with even more dumbed-down humans trotting gamely behind them—the agents are going to have to convince us that they understand our needs, if not better than our real-world friends and family do, then at least competitively so.

But how are our agents going to get that smart? Lanier thinks they won't, not ever. The humans will just get more stupid, in a kind of regressive coevolution. Tastes, after all, don't translate easily into simple formulas. Think about the resolution, the texture, of your taste in music and books: certain broad generic categories may be useful in describing your appetites (you tend to like jazz more than classical, fiction more than nonfiction), and you certainly have lists of favorite and not-so-favorite artists. But when you zoom in on that possibility space, the generalizations grow less and less useful: you might be a huge Dickens fan, but that doesn't necessarily make you a lover of Victorian serial novels written by men. GenXers might think of the gap that separates Nirvana from Pearl Jam. Ultra-hipsters retain a certain grudging respect for Kurt Cobain's band, despite its popular success, while Pearl Jam is seen as a more commercial derivation, a wannabe Nirvana with great marketing. But try explaining that distinction to the computer.

You could get as specific as you wanted, as in the statement: "I like Seattle-based, former indie, all-male grunge bands that came into popularity in 1991, with a history of drug problems." Even if you could somehow get the agent to understand all those variables, it still wouldn't be able to pick up the distinction between Pearl Jam and Nirvana, a distinction that is self-evident to anyone who follows popular music.

There are two potential consequences of this limitation. Either our tastes will increasingly conform to the cartoon versions of their previous selves (as in Lanier's model), or we'll find those cartoon versions to be excessively cartoonlike and ignore them. My hunch is that the latter will happen, but only because there will be better agents on the market. The agents that we tutor directly, by describing our tastes as a series of generic axioms ("I like best-selling detective stories written by women"), will look crude and uncultured next to the more worldly and more discriminating "social agents." Working versions of these social agents already exist, most famously on Patti Maes's Firefly Web site. If Firefly turn out to be the model for intelligent-agent technology, important shifts in our cultural appetites may well be on the way, but they won't look like the gloomy, sedated future of Lanier's bumblebees. If anything, the future according to Firefly looks to be more chaotic than the dulled, by-the-book predictability of most mass media. But to understand that potential volatility, we need first to understand how social agents work.

Firefly began as a simple program—almost a curiosity piece, really—designed to recommend music. The program would toss out a few album titles and ask you to rate each of them on a scale of one to seven. That was the input process in its entirety—no descriptions of different genres, preferred time periods, regional inflections. No long introspective essays on the evolution of your musical interests. Just a random list of records and a simple numerical evaluation. The trick was to rank as many records as possible; you'd pass judgment on ten albums, and Firefly would promptly serve up another ten. (One reason the program caught on was that this whole rating process was more than a little hypnotic. You invariably ended up with college dorm–style questions to ponder: If *Abbey Road* is a seven, what's *Rubber Soul?* Do I really like *Saturday Night Fever,* or is that just the irony talking?) Eventually, the rating cycle would grow tiresome, and you'd instruct Firefly to recommend some music in return. The original software included a button that cheered: "Go agent!" You clicked on it, and the magic began.

What the agent brought back was a list of records that you hadn't yet ranked but that it nonetheless thought you might like. For most of the people who tried the service in that first year, the whole exchange was electrifying, and more than a little uncanny. The agent would inevitably return with four or five records that were absolute all-time favorites of yours, as well as a few others you'd always meant to check out. Every now and then a dud wandered into the mix, a mismatch. (When I first tried Firefly, it kept insisting that I was secretly enamored of the speed metal band Anthrax.) But the overall effect was like a blast from the future: with only a slender, numerical profile of twenty or thirty albums, the computer had somehow built a mental model of your interests and projected out from there. It was more than a little eerie, this computer listening in on your musical appetites—that old threshold of mechanical desire crossed once again, like the calamitous ball at the end of "The Sand-Man," or Harrison

Ford's romance with Sean Young in *Blade Runner*. There was something strangely intimate about that list of records. Half the time, it felt like your ordinary high-tech wizardry. But the other half felt like a haunting.

The wizardry behind Firefly goes by the name "collaborative filtering," but its underlying premise is just common sense. Firefly relies on the transferability of taste: it assumes that people who have some interests in common will also share other interests. It's a kind of aesthetic transitivity: if you and I both enjoy Cannonball Adderly, and you also like John Coltrane, then it's more likely than not that I'll be a Coltrane fan as well. The Firefly agent is actually a distant cousin of Apple's V-Twin search engine from the last chapter. It looks for patterns—in this case the patterns of likes and dislikes that make up our musical leanings. When it finds a match (built out of dozens of rankings, not just one), it extrapolates out from the albums that happen to overlap in our respective rankings. Pink Floyd fans tend to be devotees of the mid-seventies Grateful Dead; if you happen to like *Dark Side of the Moon*, then Firefly will assume that the Dead's *Terrapin Station* is right up your alley. Firefly looks for "neighbors" in the possibility space of musical discrimination. If your tastes put you near a Prokofiev fan, but you haven't ranked the composer, chances are you'll want to check out his *Romeo and Juliet;* if your tastes "live" far from an Anthrax fan, chances are you won't respond well to other speed metal bands.

These projections, however, are only part of the Firefly magic. What makes the system truly powerful is the feedback mechanism built into the agent. In the first rounds of recommending, the agent invariably serves up records you've heard before. Firefly encourages you to rate those albums, plugging information right back into the database, and allowing the agent to evaluate the success of its picks. After I shot down three or four Anthrax albums, the Firefly agent got the hint and stopped recommending them. It had assumed from my initial ratings—which included high marks for Sonic Youth and Mudhoney—that I generally responded well to distortion-saturated bands that had a strong following in the skate-punk community, and so it kept hammering me with Anthrax. But my feedback made it clear that I had a different sensibility: the noisier bands I liked had their roots in more experimental, Velvet Underground–like traditions; the full-throttle head-bangers weren't my cup of tea.

Like the textual interfaces of the previous chapter, the Firefly agent could see that quality of my musical taste without being able to describe it or abstract out to the larger categories it represented. Dealing only with patterns— in this case, the patterns of likes and dislikes among thousands of people—the agent could make subtle distinctions between different appetites, distinctions that in some cases might not be readily apparent to those of us dutifully plugging in the numbers. (Part of the fun of using Firefly was trying to figure out what strange assemblage of music lovers you belonged to.) The number crunching behind that aesthetic judgment relies on two things to do its work. First, the larger groups of preferences are built from the bottom up, not the top down. You don't tell the computer in advance: "There are grunge fans, hip-hop fans, slo-core fans, and techno fans. Try to fit each new user into the right category." The computer just churns through the millions of ratings in its database and looks for patterns that repeat themselves. Those patterns become the basis for the different congregations of taste, instead of the fixed, preordained categories of genre or period. The second ingredient is the feed-back loop of rating the agent's recommendations. The input from the user allows the system to refine itself, to adapt to new information as it comes in. The more information in the data-base, the more feedback from the users, the smarter the agent gets. The bottom-up organization lets the computer perceive relationships that might otherwise fly below the radar of human perception. The feedback loop lets it learn.

We know from chaos math and Jimi Hendrix that feedback systems are notoriously unpredictable. The mon-strous fractal landscapes unleashed by Mandelbrot were the mathematical products of a feedback loop, as were the disso-nant swirls of "Purple Haze." The much-analyzed turbulence of the 1992 and 1994 elections, oscillating wildly between Demo-cratic and Republican landslides, was in part a product of the feedback loops of modern media coverage. The "angry elec-torate" is a creature of its own public opinion surveys, each expression of popular outrage fueling more outrage, as each poll digests and amplifies the results of the preceding poll. A State of the Union address produces an instantaneous "public opinion" result that is promptly digested by the viewing public, and then channeled back into the following day's polls, shaping the next day's polls, and so on. As the feedback becomes increasingly rapid-fire and incessant, the results become more and more chaotic, subject to dramatic and volatile shifts in intensity. The stable, linear "common ground" of conventional wisdom becomes a kind of electoral Mandelbrot set, fluctu-ating violently between opposing extremes.

If Firefly's social agents come to dominate a large swath of popular taste, we can expect the basic rhythms and shapes of cultural production to change dramatically, becoming increasingly volatile, increasingly hard to predict. We will migrate from the stultifying but stable system of mass media to the more anarchic realm of cultural feedback loops. That change is bound to be an exciting one, and perhaps even more democratic than the current system of mass-media indoctrination. But that dizzying volatility will surely come at a cost.

We can only speculate about the fallout from this new feedback-driven system. (Unpredictability, after all, is one of its defining characteristics.) But let's take a plausible extension of Firefly's current modus operandi as a case study of sorts. We might as well use the holy grail of all couch surfers and high-tech trendspotters: a customized version of MTV— programmed by your Firefly agent instead of the Viacom upper management. We're not talking anything fancy here—no Aerosmith-style, star-in-your-own-video pipe dreams—just a straightforward setup, the sort of thing that you could almost rig together today using a Touch-Tone phone as the feedback device. The system would work much as Firefly does today: MTV serves up ten or fifteen random videos and you rate each one, using the familiar one-to-seven scale. After a batch of random selections, MTV starts trotting out the videos it thinks you'll groove with—like an inept suitor on a first date, shuffling wildly through his record collection, looking for something that'll set the mood. The first offerings will be hit-or-miss, but you'll be able to train the agent through the feedback mechanism. As your private MTV grows more in sync with your sensibilities, millions of other viewers train their personal veejays as well. That collective, bottom-up process generates a tremendous number of potential patterns to churn through, and the viewer feedback helps the agent figure out which patterns are worth keeping. After a few iterations, you have a customized

MTV smart enough to anticipate which new videos will suit your tastes, while still rotating steadily through your old favorites. It might not be a perfect system, but it'd certainly be more entertaining than the present incarnation.

Like the Web version of Firefly, this agent-driven MTV would be vulnerable to Lanier's counteragents—mainly in the form of automated programs designed to stuff the ballot box in favor of specific artists. But MTV and Firefly could easily rig together some sort of authentication system to verify that a *person* was rating the videos (and not a 'bot); the only way to hack the system after that would be the decidedly low-tech route of paying thousands of screenagers to skew the database toward certain artists. As it turns out, the problems wouldn't arise from people *outside* the system trying to commandeer its pattern landscapes; the real problems would come from the internal volatility of the system. That volatility is bound to provoke varying responses from the public experiencing it. Some will welcome it into the world, the way the Futurists embraced the blistering human-machine hybrids of the race car and the howitzer. Others will lament the passing of the older ways, even if they were less "interactive." But where would all this turbulence come from in the first place?

Think about the way contemporary tastes in popular music evolve. For the past three or four decades, the pop music world has been divided into two ecosystems: the top-down, one-to-many, mass-mediated system of Billboard talent; and the bottom-up, many-to-many system of local artists, subcultures, avant-gardes, and so on. On the one side, popular tastes are shaped largely from above—conjured up out of marketing blitzes, cross-promotions, television appearances. The record executives decide to "launch" a certain band, and after a relentless campaign of numbing repetition, the general public finally says uncle and learns to sing along. On the subcultural side, artists float in and out of different audiences, as word passes from mouth to mouth. There are more artists in the mix—no single band comes to dominate the subcultural sphere—but then again, there's little mass acceptance. The resources are more equally distributed, but there are fewer goods to go around.

The Firefly model does away with the old Manichean opposition between mass culture and subculture. For one thing, there's no executive branch dictating which artists get the promotional campaigns and airplay. And even the big-budget commercial bands are subjected to the many-to-many, word-of-mouth evaluations that once characterized the subculture. At first glance, it seems like an ideal win-win scenario, a democratic revolution in popular taste. But the capacity for self-organization should not be tampered with lightly. It has a way of biting back.

The divided governance of today's pop music universe actually has its benefits, if only because there are conduits that flow from indie labels to the majors, conduits that are for the most part maintained by MTV (assisted by the more open-ended playlists of college radio). The two systems complement each other. With their marketing muscle and the sheer force of repetition, the major labels create sustainable, "household-name" artists who continue to produce successful records decades after their initial luster has worn off. But the mainstream music world is a conservative system, in the most basic sense of the word: it reinforces the existing state of affairs. Record executives, as a rule, prefer successful bands to less successful ones. Subcultures, on the other hand, are much more receptive to innovation. They value musical originality over a proven, commercial track record. Indie bands attract audiences because they sound *new;* mainstream bands attract audiences because they've already attracted them. But the subculture invariably feeds into the dominant culture, and it is that steady trickle of innovation that keeps the mainstream lively. The record execs are still controlling the playlists, but they're tossing in new variables every few months, based on the latest happenings in the subculture. The result is often deplorable (think Faith No More or Vanilla Ice), but it has a certain best-of-both-worlds feel to it. You get the stability of mainstream culture with the innovation of the indies.

How would the world of popular music change, were MTV programmed entirely by intelligent agents? For a start, the intricate back-and-forth between the mainstream and its tributaries would subside. In its current form, MTV mediates between the two spheres of pop music: the videos are handpicked by the programmers, but the programmers are trained to be responsive to the latest word on the street. Firefly's agents would change all that. Feedback systems tend to gravitate toward certain hot points, what the mathematicians call "attractors." They have a tendency to loop back into themselves, create self-fulfilling prophecies, as in the "angry electorate" example. A poll finds that 60 percent of the electorate feels that candidate X is corrupt; the next day a certain fraction of the electorate begins to associate the candidate with shady political dealings, having seen the results of the poll. Their new attitude is reflected in a new poll, and so on ad infinitum. Over the next few months, the numbers continue to climb steadily, driven more by previous polls than by actual news. What makes this system unstable is the feedback mechanism: if the pollsters were contacting people who were sequestered from the poll results, you'd have a steadier, more reliable public opinion.

You can already see this phenomenon at work on Firefly, in what I call the Beatles-and-Bach syndrome. One of the major complaints about Firefly—particularly from sophisticated music lovers—is the system's propensity for low-risk, low-information recommendations. It's not terribly rewarding to sit there rating albums for an hour and then have the agent come back with: "I think you'd like *Sgt. Pepper's Lonely Hearts Club Band* and the *Goldberg Variations.*" The evaluation may be accurate enough—you happen to like both records—but the information is of no value to you. Not just because you already own the records, but because *everyone* likes the Beatles and Bach. You don't need an intelligent agent to tell you this. What you want the agent to do is find those obscure records that you've never heard but that match your tastes perfectly. The Firefly agent is ill-inclined to do that, though, because there's so much data reinforcing the *Sgt. Pepper's* match. Think about it from the hapless agent's point of view: almost every time it offers up a Beatles disk as a recommendation, the user responds favorably. (The agent, of course, can't tell that the user is muttering at the screen, "Yes, I know, the Beatles. Tell me something I don't know already.") All the agent wants to do is make an educated guess and be rewarded with a high rating. Like a well-trained dog, it quickly learns that recommending the Fab Four is the most dependable way to get a positive response from the user, and so the agent starts relying on the Beatles as a crutch, a sure thing.

There are ways around this particular problem— some of which Firefly has already implemented in the latest version of its software. Just as Apple's V-Twin engine discarded low-information words like *the* and *and,* the Firefly agent can discard, or at least deemphasize, low-information artists—musicians that everyone either likes or hates. But the phenomenon itself is a telling one. The idea of a broad, across-the-board consensus—or even familiarity—with a single artist belongs very much to the mass-media paradigm, where the promotional budgets have anchored certain musicians in the popular imagination. Top-down systems tend to trudge through the same old favorites, dictating that everyone listen to a smaller, more predictable playlist of artists. That's one major reason that we have universally revered bands in the first place. The Beatles-and-Bach syndrome is a kind of historical overlap, a discontinuity: it's a self-organizing system trying to evaluate tastes that were formed largely under a more static, hierarchical regime. But what if the music world were shaped entirely by bottom-up systems like Firefly? What would happen to our tastes then?

Certainly they would diversify. The usual conventions of standardized culture—the pop song that everyone knows by heart—would grow less and less common; the music world would subdivide into smaller niches, built out of the pattern matches of the Firefly software. Those clusters would tend to reinforce themselves, making "crossover" hits even more rare than they are today. The clusters would be hard to predict, and even harder to describe, but they would exert a strong gravitational pull on new listeners. Within those clusters, individual artists would surge and subside at the unsteady rhythms of today's public opinions polls. A new band that happened to attract a few core listeners—users whose patterns synchronize well with the tendencies of the overall group—might shoot to the top of the playlists much faster than the latest pop phenomenon does today. These self-fulfilling attractors could arise out of the slightest statistical anomalies: no laboring away for years in dive bars and opening acts, honing your craft until a label scout takes notice. A band that stumbles across the right listeners could skyrocket into major following overnight. By the same token, bands could plummet into obscurity just as quickly. Feedback systems tend to work that way. They spiral out toward various extremes, for no apparent reason. They have no ear for history or tradition. If the music business comes to rely increasingly on social agents to promote its records, that self-reinforced turbulence will increasingly define what we like and what we don't like in our music.

This may be the great irony of the intelligent agent. While its critics prophesy for us a dulled, lifeless future enslaved to the agent's prerogatives, the real legacy may turn out to be one of turbulence and unpredictability, our old cultural appetites fractalized, torn asunder. On this front, of course, the cyborg literature has only a bleak prognosis to offer. The human-machine hybrids of *Blade Runner* and *Frankenstein* came into the world as simplifiers, problem solvers, labor savers. But by the end they were chaos machines. We can only hope the true cyborgs of the future turn out to be more stable than the fiction.

INFINITY
IMAGINED

Anyone who followed the "canon wars" of the late eighties knows that history departments and literature seminars have been experimenting with a new form of historiography in recent years: subaltern studies, people's histories, "archaeologies" of the past—all told from the viewpoint of groups usually silenced in traditional accounts. The cast of characters is familiar by now: factory laborers, the mentally ill, Native Americans, working women, criminals, gays and lesbians—the whole parade of "otherness" championed by progressive thinkers and ridiculed by back-to-the-canon conservatives like William Bennett and Dinesh D'Souza. Maybe it's time to add to that list the ever-growing dustheap of obsolete technologies—not only the machines that were outmoded by sleeker or better-marketed competitors, but also the machines that never found a market at all, despite possessing superior technology. If our social history now belongs to the outcasts and the oppressed, then perhaps our high-tech history is due for the same reversal of fortune.

What would this kind of history look like? For the most part, it would be dominated by cranks and tinkerers, the sort who file for a hundred patents in a lifetime and never make a penny for their labors. The armchair inventor and the gadget freak didn't preside over the lore of nineteenth-century capitalism the way the dashing young man about town did, but the figure of the neighborhood tinkerer certainly had its moments. Most of us remember the ending of *Madame Bovary* for the heroine's suicide at the hands of pulp fiction, but the novel actually ends with the pharmacist Hommais—the amateur inventor and full-time crank—being awarded the Prix d'Honneur. You can see this denouement as an emblem of Flaubert's dark irony, or his obsession with the pathologies of "modern stupidity." (What better antihero for *Madame Bovary,* a book about the illusions of the mass-marketed romance, than the bothersome, know-it-all next-door neighbor?) But you can also see in that ending a remarkable prescience. If Emma Bovary is the great literary ancestor of the modern tabloid-addled suburban housewife, Hommais belongs uniquely to his own era—all those gentlemen of leisure concocting new mechanical plowshares, or those precocious twenty-somethings experimenting with metallurgy and magnetism in sooty, lamp-lit chambers.

Seen from the right angle, Hommais turns out to be nothing less than an incompetent, Gallic rendition of Edison—the definitive icon of late-nineteenth-century entrepreneurial capitalism. And even Thomas Alva himself would play a prominent role in our alternative history of obsolete machines. The man may have powered up the first electricity grid and sung "Mary Had a Little Lamb" into the first phonograph, but the labs at Menlo Park and West Orange also produced a steady stream of duds alongside the success stories. This is to be expected, of course: significant inventions are like omelettes—you have to break some eggs to make them happen. But we would do well to spend some time contemplating those broken shells, to learn more about the discards and miscarriages, the "creative destruction" that propels all high-tech advancement.

You could build an entire academic press around this line of inquiry, though the end result might look more like a machinic freak show than a serious body of research. In the end, I suspect, the most interesting studies would gravitate not toward the cranks and useless gadgetry, but rather toward the legitimate visionaries and their breakthrough inventions—if only to show how hard it is to project outward from a major technological advance, to see beyond the mechanical details to the machine's broader social consequence. Remember Edison's description of the phonograph and its future applications? How many of those had anything to do with what phonographic technology eventually became useful for? Only a small fraction, of course—because technohistory is littered with unintended consequences and limited fields of vision. According to Edison, the record player was an upgrade for the telephone medium, an enhancement. Perhaps people would occasionally listen to prerecorded music on the device, but for the most part, Edison thought, they'd be sending each other dictated letters through the postal system—like today's voice mail without the immediacy.

As always, "the street finds new uses for things." What's remarkable here is not that the street appropriates the technology but that we have such a hard time envisioning those appropriations before they happen. It's as though the sharp, luminous shock of revelation—the eureka moment of all inventor mythology—carries with it a certain haziness, a glare that blocks out as much information as it reveals. You stumble across a way to record voices, but you can't see what it's good for. You predict the rise of the desktop PC, but all you can imagine it doing is filing cooking recipes. This is the hard bargain of life on the cusp of high-tech paradigm shifts: you're blessed with a certain technical enlightenment, but it's difficult to see much beyond that bright knowingness. Blindness and insight—you can't have one without a solid dose of the other.

Nothing illustrates this point more powerfully than Vannevar Bush's wondrous Memex device, now widely considered one of the PC's venerable ancestors. As we saw in chapter 4, Bush's speculations on the associative powers of the Memex anticipated much of the modern PC's storage-and-retrieval capabilities, as well as the hyperlinks of the World Wide Web. But this is a selective, hindsight-driven reading of "As We May Think," one that emphasizes the passages where Bush gets the future right and ignores the many sections where his vision is decidedly less clairvoyant. You can describe the Memex as an information processor that enables you to store old documents, write notes to yourself, organize data, and perform calculations—all while sitting at your desktop. That sounds a great deal like your everyday PC circa 1997. But you can also describe the Memex in a different fashion, punching up other elements in the mix, elements that seem less consequential to us now because they didn't come to pass. Consider just this short inventory:

1. The user captures information via a small camera lodged on her forehead or her glasses, snapping pictures of documents as she reads them.
2. The documents are transformed by "dry photography" into small microfilm-style images, which are stored in the body of the Memex device.

3. The storage mechanism is a linear roll moving from left to right, with each frame on the roll containing thousands of miniature documents.
4. When the user is away from her desktop, she enters text into the device using spoken words, transmitted via radio communication.

The list could go on. (Bush spends several pages at the essay's outset exploring the technical possibilities of dry photography, a reproduction method that has absolutely nothing to do with the modern computer.) In each instance, the machine described appears to be a completely different species from the modern desktop PC—more like a souped-up microfilm device that has been crossbred with a photocopier. Some might consider this a matter of quibbling over minor details. So what if Bush didn't anticipate the microprocessor or the video monitor? Surely it's enough that he came up with the basic vision of a desktop information processor. After all, no one else at the time had managed such a remarkable imaginative leap. Who cares if he happened to be distracted by the red herrings of dry photography?

These objections might be more persuasive if the Memex's dry-photography foundations didn't have such profound consequences for the device itself. The photographic medium is static, immutable—you take a snapshot of a page and it's frozen in that form forever. In Bush's system, even the notes entered directly by the user were captured on microfilm and remained crystallized in that original state for the rest of their existence. You could "interact" with documents by linking them to other documents using Bush's brilliant system of "trails"—but you couldn't actually *edit* them, change words around, add paragraphs, delete whole passages. You could organize documents using the powerful associative tools that Bush conjured up, but you couldn't manipulate their contents. This was no minor oversight. The ability to alter the content of a document—experimenting with different phrasings, rearranging things, cutting and pasting—this may be the defining characteristic of the digital computer, what separates it from its mechanical predecessors. Imagine a word processor or a spreadsheet that let you enter one draft of a document and then prohibited any subsequent alteration—it would be an appalling product, of course, but it would also suggest a fundamental misunderstanding of the digital medium, like an oven without a temperature knob or a radio tuned perpetually to one station. The power of manipulation is the sine qua non of the modern computer, its core competency. And Vannevar Bush missed it altogether.

I bring up this point not to take anything away from Bush's prophetic essay, but to introduce a larger argument about our own historical moment and this strange new medium of interface design. Bush, of course, described more of late-twentieth-century technology in that short essay than anyone before him—and for that he deserves pride of place in the annals of digital computing. (Indeed, as I tried to show in the "Links" chapter, today's interface designers would do well to be *more* faithful to certain elements of the Memex's architecture.) But for all his extraordinary insight, Bush couldn't see the PC's defining characteristic, the malleability of digital information. There is a lesson here for anyone who attempts to make sense of the high-tech world, a lesson that is close to the heart of this book's primary thesis. At the threshold points near the birth of new technology, all types of distortions and misunderstandings are bound to appear—misunderstandings not only of how the machines actually work but also of more subtle matters: what realm of experience the new technologies belong to, what values they perpetuate, where their more indirect effects will take place.

Twenty years ago, the graphic interface seemed like a toy, virtual training wheels for computer novices. Now we readily accept it as a necessity for serious computing: functional and easy to use; an essential tool for power users and neophytes alike. But to go beyond that efficiency model and see the graphic interface as a medium as complex and vital as the novel or the cathedral or the cinema—that's an assumption that still requires some getting used to. The recent battles between techno-utopians and neo-Luddites have not helped matters much. One side announces that the Internet is the "greatest invention since the discovery of fire" while the other eulogizes the death of the slower, more introspective consciousness of print media. The cultural impact of new technology is hard enough to predict without the fury of manifestos obscuring our view. This, in fact, may be the most important lesson to draw from "As We May Think": not the dead-on predictions or the false leads, but instead the tone of the essay itself, which is sober, reflective, exploratory, intent neither on burying the past nor on renouncing the future. Vannevar Bush may have neglected a few critical elements of the modern PC, but the general sensibility of his prose should be a model for all techno-criticism to come.

What, then, are the blind spots of our own age? We have already encountered a few: the tyranny of image over text, the limitations of the desktop metaphor, the potential chaos of intelligent agents. But there is a more fundamental—and for that reason more difficult to perceive—blind spot in the high-tech imagination, and it has to do with the general region of experience that the interface is felt to occupy. Until very recently, interface design belonged squarely to the geeks and computer hobbyists—a niche market at best. The rise of the Mac and Windows introduced a mass audience to desktops and icons, while the Web's popularity endowed browsers and hypertext with a certain subcultural sexiness. All these developments suggest a widening of the interface audience, but the medium itself still belongs to the world of functionality and increased convenience. We're subjected to endless advertisements promising us a miraculous digital future, and yet the scenarios they deliver tend to be remarkably mundane: ordering concert tickets, reviewing X rays from a remote location, sending photos to relatives by e-mail. There is a strange mix of narrowness and wild boosterism in this climate: we're reminded a dozen times each day that the digital revolution will change everything, and yet when we probe deeper to find out what exactly will change under this new regime, all we get are banal reveries of sending faxes from the beach.

The most profound change ushered in by the digital revolution will not involve bells and whistles or new programming tricks. It will not come in the form of a 3-D Web browser or voice recognition or artificial intelligence. The most profound change will lie with our *generic* expectations about the interface itself. We will come to think of interface design as a kind of art form—perhaps *the* art form of the next century. And with that broader shift will come hundreds of corollary effects, effects that trickle down into a broad cross section of everyday life, altering our storytelling appetites, our sense of physical space, our taste in music, the design of our cities. Many of these changes will be too subtle or gradual for most people to notice—or rather, we'll notice the changes but we won't perceive their relationship to the interface, because the various elements will appear to belong to different categories, like so many aisles in a grocery store. But the history of technoculture is the history of such interminglings, the unlikely secondary effects of new machines rippling out to transform the society that surrounds them.

The most fertile historical analogy for this process is the invention of perspective in painting. When Brunelleschi and Alberti hit upon a way to create the illusion of depth on a two-dimensional surface in the early fifteenth century, you could see their techniques—the vanishing point, the picture plane— as just another clever trompe de l'oeil, a curiosity piece. Certainly, it was an improvement on the muddled visual space of medieval art, but artists were always coming up with new techniques to advance their craft: chiaroscuro, the camera obscura, pointillism. Perspective, however, turned out to be more than just a minor enhancement to the painter's repertoire. The mathematical studies of Alberti and Leonardo transformed not just the spatial language of European painting but also the role of the artist itself, elevating painting to a higher cognitive stature—closer to science or philosophy than to popular entertainment, and in doing so helped create the whole notion of the artist as intellectual. Perhaps more important, perspective centered the visual field on the human point of view, instead of a disembodied or divine locus, a shift that was imitated in countless disciplines throughout the fourteenth and fifteenth centuries as scholars and artists and scientists grounded their work in the physical, lived reality of the human body. Perspective began as a technical innovation, but it eventually helped produce what we now call the Renaissance.

The discovery of information-space may engender a social transformation as broad and as variegated as the one that followed Alberti's marvelous breakthrough. And that is why it is so essential that we acknowledge the medium's richness and complexity, its range of expression and its cultural import. Every major technological age attracts a certain dominant artistic form: the mathematical and optical innovations of the Renaissance were best realized in the geometry of perspective painting; the industrial age worked through its social crises in the triple-decker novel. This digital age belongs to the graphic interface, and it is time for us to recognize the imaginative work that went into that creation, and prepare ourselves for the imaginative breakthroughs to come. Information-space is the great symbolic accomplishment of our era. We will spend the next few decades coming to terms with it.

In the end, this book is only a preliminary survey of the field, a glimpse of the new medium in its formative years as it gropes uneasily for new ways to represent information. We can look forward to a great deal of maturation in our interfaces over the next few years. A decade from now the desktop metaphor may seem as quaint and bewildering to us as the command-line interface does now. On the other hand, certain interface elements may remain constant over time: the window, for instance, appears to have a certain durability—not unlike the Baroque frames that survived several generations of artistic fashion in the seventeenth and eighteenth centuries. Part of the point of this book, of course, is that we can't always predict what will change and what won't— that's one reason that the technology is so powerful. What is clear, however, is that the influence of this technology will extend well beyond the traditional scope of the computer interface, just as Renaissance perspective transformed more than the frescoes and basilicas of Florence and Rome. I have tried to sketch some of those unlikely side effects and migrations in the preceding pages. The reader may judge whether they seem plausible, and history, of course, will grant us the final account.

While the details may change as the form evolves, I think it is possible to settle on a few broad themes or tensions in the interface medium—themes that will come to dominate both highbrow and lowbrow experiences of information-space over the next decade. I suspect some readers will already have detected other themes winding through the preceding pages; for clarity's sake, I have tried to deal only with the major threads here, the ones poised to dominate the field for some time. In other media, of course, such thematic oppositions are commonplace, and most of them end up outlasting the artistic movements that first brought them to the forefront of debate. The novel, for instance, has been wrestling with the demons of psychological depth ever since George Eliot and Henry James began to explore the full dimensions of late-Victorian mental life. (D. H. Lawrence once said that Eliot was the first to write novels where the most important events took place in the characters' heads.) The battle between introspection and social portraiture lies at the very heart of modernism, of course, and even extends to the vacant, brand-saturated wilderness of the K mart realists and other postmodern writers. At least a century of novel writing has agitated over that divide—so much so that the tension between inside and outside in modern fiction almost goes without saying now, a received idea last discussed in earnest during high-school American lit. But the theme itself still exerts an enormous influence over the way that we make sense of the novel as a form.

The problem with the interface medium at present—and this is one reason that we have trouble taking it seriously *as* a medium—is that we don't have a language like this to describe it. For the most part, our evaluative criteria reduce to the bottom-dollar question: is it easy to use or not? There's invariably a bonus round for the cyber-slackers—is it cool?—but that's usually where the critique comes to a grinding halt. As I've tried to show in the preceding chapters, it's not that our interfaces are lacking in imaginative depth or complexity; it's just that we don't have the critical vocabulary to deal with them in anything but the most rudimentary terms. What follows is an attempt to sketch out a few major oppositions that will hold sway over the interface medium for at least the next ten years. Imagine these themes as templates of sorts, to be filled out by the detail work of countless interface artists to come. We need their labors and their insights to grasp the emerging stature of the interface medium, to see it in its full glory. As Eliot wrote in *The Mill on the Floss,* "The full sense of the present could only be imparted gradually by new experience—not by mere words which must remain weaker than the impressions left by the old experience."

Spatial Depth Versus Psychological Depth

In the spring of 1994, Broderbund Software released Robyn and Rand Miller's classic interactive adventure Myst, the gaming world's elegant and vaguely Borgesian ambassador to highbrow culture. Quickly dubbed the first "video game for adults" and the "*Ulysses* of CD-ROM," Myst attracted the kind of contemplative, sober analysis usually reserved for art films and literary biographies. The Miller brothers themselves seemed headed for certifiable cult status, auteurs for the digital age, a hybrid of David Lynch and J. R. R. Tolkien. But the hype led quickly to the inevitable backlash, and in November of that year, the *Washington Post* ran a long story by its Pulitzer Prize–winning critic Michael Dirda that took issue with the game's inflated artistic reputation. As entertainment, Dirda argued, Myst offers a mixed bag: mediocre game play at a sedentary pace set against a lavish, fully-rendered backdrop. As a work of art, however, Myst didn't make the grade: "The characters are ciphers," Dirda wrote, "the language nearly nonexistent and the plot trite." In other words, gamers looking for stunning graphics will be well accommodated on Myst's lavish isle, but aesthetes hankering for cutting-edge, digital art forms shouldn't purchase that CD-ROM drive just yet, given Myst's limited offerings. If you're looking for psychological depth and literary complexity, Dirda suggested, you're still better off with the analog pleasures of Henry James and William Faulkner.

Dirda had a point, of course—the characters in Myst were as flimsy and low-resolution as the digitized clips they appeared in, and at the rare points where the writing was halfway decent, the lines were invariably mangled by the obligatory faux English accents of all CD-ROM acting. If this was supposed to be a *Ulysses* conjured up out of zeros and ones, then where was the cognitive depth of Joyce's novel, the ambulatory and absentminded central intelligence of Leopold Bloom or Stephen Dedalus? It was tempting to see in Dirda's critique an echo of Sven Birkerts's eulogy to the deep consciousness of the traditional novel:

> As our culture is rapidly becoming electronic, we are less and less what we were, a society of isolated individuals. We are hurrying to get on-line, and the natural corollary to this is that the idea of individuality must come under siege.... In time

we will all live, at least partially, inside a kind of network consciousness. . . . Our spells of unbroken subjective immersion will become rarer and rarer, and may even vanish altogether.

Reading Dirda's review alongside this passage, you wonder whether the flattening out of experience that Birkerts describes has met its symbolic match in the thin, undeveloped characters of Myst. We get the narratives we deserve, after all. If the hive consciousness of global networking has done away with "subjective immersion," then it's no wonder we're satisfied with the empty mental life of the Miller brothers' creation. We don't notice the limitations of the art because our own sense of self has been whittled away by the dark forces of perpetual connectedness.

This sounds like a compelling reading, but it is predicated on false assumptions. Like so much of contemporary interface design, Myst is primarily a spatial experience. If there is immersion, it is the immersion of locale, the strangely hypnotic feeling of exploring a terra incognita, of losing your bearings and then finding them again. The aesthetic pleasure of Myst is closer to the environmental jazz of certain architectural projects, where chance and disorientation are an explicit part of the package—environments like the Parc Villette installation outside Paris, or the eclectic sculptures scattered throughout Manhattan's Hudson River Park. (The lowbrow equivalent of all this, of course, is the densely imagineered rides of Disney World.) If there are no lifelike characters in Myst's fictional world, that is because the world itself is more important that the characters that populate it. Denouncing Myst for its lack of character development is like finding fault with an office building for its lack of emotional sophistication. We don't expect our architects to re-create the intensities of human consciousness—why should we expect anything more from our interface designers?

You can see this distinction most clearly in the popular forms of recent video games, like Sega's Sonic the Hedgehog franchise. The visual iconography and storylines of these games are literally cartoons, pitched at a ten-year-old's level of sophistication. Widely hyped as the "fastest game on the planet" when it was released in the early nineties, Sonic wasted almost no time with complicated puzzles or thumb-straining feats of manual dexterity. For the most part, you blindly whizzed along, scrolling at high speeds past a luminous back-drop, bouncing and plummeting and catapulting along the way. For all the kinesis, the hapless Sonic addict had little control over the onscreen character's actions; there were really only two options—jump and go faster—and pretty much any combination of those two would produce something interesting on the screen. The lack of control wasn't perceived as a drawback because the whole point of the game—what made it such a phenomenal success—lay in the sheer exhilaration of moving, and moving *fast*. You didn't so much play Sonic as ride it. Its genetic code was closer to a roller coaster than to a board game. Sure, there were levels you could advance to, and the occasional trap-door or secret passageway, but these were largely vestigial elements, left over from the conventions of the game's more sedentary predecessors. The game was finally all about the rush and the intensity of moving through digital space; you didn't need puzzles or plotlines for that. The game captivated its audience for environmental reasons, not narrative ones. Subsequent blockbuster games—like Nintendo's lush, 3-D-rendered Mario

64—were merely variations on Sonic's original theme, performed with more advanced instrumentation. The space was what mattered. Everything else was incidental.

For the neo-Luddites, of course, this hardly satisfies as a defense of the medium. Even if you give up on the idea of psychological depth, surely there's something oppressive in the mindless acceleration of Sonic the Hedgehog and his ilk. An art form predicated on speed alone is bound to remain at the aesthetic level of roller coasters and amphetamines. As with so much of today's techno-commentary, the critique is half-right. If we were doomed exclusively to a succession of Sonic imitators, our future would indeed look bleak. But the information-spaces of Sega and Nintendo are only leading indicators in this field, a glimpse of the future conveyed to us by the modest means of the present. The audiences that roared along with *A Trip to the Moon* —Georges Méliès's 1902 special-effects extravaganza—could sense that something potent was in the works, but the idea that those jittery, flickering images would somehow evolve into *Citizen Kane* and *Vertigo*—or even *Jurassic Park*—would have seemed preposterous. Sonic and Mario are the precocious infants lying at the base of what will become a formidable family line. We can't predict what their descendants will look like, but we *can* be sure that the exploratory, spatial quality of the medium—the *haptics* of information-space—will be of enormous importance to that tradition, whatever it turns out to be.

Society Versus the Individual

All great symbolic forms address the conflict between the private self and the larger community that frames that self, whether this valuation lies at the surface of the work or is buried somewhere in its underlying assumptions. Most architecture gravitates naturally toward larger congregations of people, just as most abstract art centers itself on private, subjective contemplation. There are exceptions, of course—I think of certain International Style skyscrapers that deadened the lively, civic interaction that had once existed on the sidewalks beneath them—but the more interesting cases tend to be those where the form itself is not hardwired to accentuate one over the other. The cinema, for instance, has a dual tradition of psychological depth and social extroversion: the wintry mindscapes of Bergman's *Persona* next to the intertwined, communal narratives of Altman's *Nashville* or *Short Cuts*. Most enduring cinematic works have been a balancing act of the two: the broad social sweep of Charles Foster Kane's publishing empire measured against the lost childhood of Rosebud; the vast international conspiracies of *Klute* countered by the stark, direct-address shots of Jane Fonda talking through her problems on the therapist's couch.

For a long time, the interface medium has concentrated most of its energies on the individual, for understandable reasons. The personal computer was just that, a *personal* computer, designed from the ground up to be used by a single individual, which is why most modern graphic interfaces draw so heavily on the imagery of desktops and closed-door offices. That symbolic sleight of hand is rightly celebrated, but who knows what imaginative avenues it closed down to us. The desktop metaphor is by definition a monadic system; it belongs to the individual psyche the way Freud's case studies do, and that inwardness can make it harder to think in more social, more communal terms. Longtime Netheads never tire of talking about the way the Internet explosion blindsided many so-called silicon soothsayers. (The first issue of *Wired* barely mentioned the Net.) Perhaps the success of the virtual desktop contributed to this myopia, a zero-sum game of sorts, where the rise in one model's fortune presupposes an equivalent, and opposite, reaction in the other. Surely thinking in the language of solitary rooms must, on some basic level, make it more difficult to think in the language of public spaces.

Interestingly, it turns out to be harder to represent communities using the tools of the modern graphic interface. There have been a number of attempts at extended metaphors: Magic Cap's 3-D office space opened onto a virtual "downtown" that represented all the user's online activities; Apple's e-World service dabbled tantalizingly with a "town square" metaphor. Both designs were hyped heavily at their launch and then quickly fizzled. The irony is that to this day, some of the most engaged and elaborate virtual communities on the planet rely on text-driven interfaces that wouldn't have looked out of place in the seventies. (Most members of ECHO and the Well still rely on command-line interfaces for their digital socializing.) This can be taken as yet another sign that the power of text is underestimated by today's reigning design orthodoxy, but it should also be seen as a call to arms for the next generation of interface designers, a genuine problem in search of a solution.

Already the VRML worldscapes and the floating orbs of The Palace suggest that new metaphors are on the way, though most such virtual spaces have the air of a product demo about them, a proof-of-concept for a concept that still needs proving. Do people really want the environmental trappings of lived space—the lavish furniture, the gothic chambers, the glittering city lights—surrounding them as they type to each other, or are these merely fringe benefits, distractions from the main event of live chat itself? The most promising recent designs have broken away from the more predictable models of town squares and watercoolers, abandoning fully realized environments for the two-dimensional, static frames of the funny pages. Microsoft has a wonderful product called ComixChat that dresses up participants with the onscreen visages of cartoon characters and scrolls through the conversation in thought balloons. There's something immediately appealing about the comic-strip metaphor, a sense of it working on the right *scale*. If the language of live chat is necessarily stripped down to the abbreviated essentials, then perhaps it's only fitting that the visual accompaniment be flattened as well. A chat-room pickup exchange that takes place in an ornate ballroom makes as much sense as spray-paint graffiti propped up against the wall of the Louvre (*pace* Jean-Michel Basquiat). Even if the visual metaphor is a compelling one, the context can overwhelm the conversation.

Mainstream Versus the Avant-garde

Nothing will propel the interface toward the status of art more quickly than the development of a functional interface subculture—small pockets of designers working in opposition to the mainstream. Coherent, self-styled avant-gardes first appeared in the metropolitan cities of eighteenth- and nineteenth-century Europe, most notably in Paris. The two worlds of subculture and mainstream have existed ever since in an uneasy, but generative relationship: the avant-garde's flair for novelty prodding the dominant culture's more conservative inclinations, a system of checks and balances that is by now so commonplace that we can barely imagine an alternative. If it is sometimes difficult to accept the artistic aspirations of the interface medium, its lack of an intelligible subculture may be at least partly to blame. For it is the condition of any nascent medium that the innovators and the establishment be indistinguishable during its formative years. (It took television nearly thirty years to cultivate a genuine avant-garde of video activists and performance artists.) Interface design has had its fair share of wayward visionaries who never made a dime off their insights (Doug Engelbart and Ted Nelson come to mind), but for the most part its major breakthroughs have been targeted at mass audiences. The system still rewards commercial success over any other potential attribute. Art for art's sake doesn't exactly open doors for you among the venture capitalists of Silicon Valley.

But the very technological advances bankrolled by those VC funds are going to change all this, and nowhere more profoundly than on the Web, where the barriers to entry are so low as to be nonexistent. In the days of Xerox PARC, you needed an entire research department to dabble in interface design, and finding an audience for your new information-space required prodigious distribution resources. On the Web, the latest visual metaphors can find their way into circulation for a tiny fraction of the cost, which means that more experimental forms—forms more interested in pushing the envelope than pleasing the masses—will naturally prosper in this environment. Much has been said about the self-publishing revolution made possible by the Web, the egalitarian dream (or nightmare) of a nation populated by millions of living-room pundits. But the real revolution unleashed by HTML may well be the democratization of interface design. The task of imagining information will no longer belong exclusively to the high priests of programming; anyone moderately comfortable with a PC will be able to concoct his or her own infoscapes, and share them with friends or colleagues. Out of this more open-ended system, a legitimate interface avant-garde will emerge. You can already see its first stirrings in the infinite-loop hyperlinks of Suck, the eclectic multimedia "installations" of Jaime Levy's design for the Web 'zine Word, and the hallucinatory VRML worlds now appearing online inspired by rave culture and the novels of William Gibson.

The rise of an interface subculture will no doubt bring a new legitimacy to the medium, at least among the connoisseurs and the curators of High Culture. But beyond the external approbation, the digital avant-garde will also bring about an intriguing reversal in the basic rules of interface design. Put simply, an interface subculture opposed to the mainstream is bound to select for information-spaces that are deliberately confusing, environments designed to perplex more than to acclimate. Just as musical subcultures confound our melodic expectations with dissonance and unusual tuning schemes, the new interfaces will strive for disorientation—or if not that, then at least new ways of orienting, so new that they confuse on first encounter. Think of the contorted, postmodern built spaces of Rem Koolhaas and Frank Gehry, buildings that appear to have been turned inside out, like Richard Rogers's design for the Centre Pompidou. It is in the nature of any avant-garde to mess with our expectations, to keep us guessing, and for the most part, we've grown comfortable—even jaded—with this endless cycle of envelope pushing. No culture in history has so readily assimilated its avant-garde movements—just look at Disney's relationship to cutting-edge architecture, or MTV's usurpation of underground video-editing techniques.

All of this suggests a reasonable blueprint for the future of interface design: the subculture spins out the innova-

•

i
n
t
e
r
f
a
c
e

c
u
l
t
u
r
e

tions, and the dominant culture appropriates the forms it thinks it can market to a mass audience. But the transition is not likely to be a smooth one, if only because the field of interface design has been governed for so long by the cardinal rule of ease of use. An information-space that deliberately disorients its occupants is bound to be dismissed for its poor design, just as the critics of Stravinsky's day fulminated against the shapeless noise of *Le Sacre du Printemps*. As a product of engineering, interface design necessarily works in the interest of clarity and coherence, but once its practitioners begin to think of themselves as artists, those values grow more and more restrictive. The first generation of interface designers to break dramatically with the first principle of navigability will no doubt be pilloried by the digital establishment, but they will also open up a whole new possibility space for the designers that come after them. The DOS snobs that turned up their noses at the Mac's desktop metaphor did so because Apple's look-and-feel seemed too easy, more like a novice's training wheels than a legitimate software advance. The interface subcultures of the future will offend the traditionalists by being too *difficult*. "User-hostile" may sound like an odd goal for interface design, but the truth is the field could use a little tough love. No medium has managed to reach the status of genuine artistry without offending some of its audience some of the time. Even under the user-friendly dictates of interface design, you can't make art without a good measure of alienation.

One Interface or Many

Interface subcultures won't go very far, of course, if their more enigmatic spaces can't eventually be conquered, made sense of.

Twelve-tone music and abstract expressionism struck many initial observers as noise and empty scribbling, but audiences eventually developed a taste for each. (Rothko derivatives now hang in the lobbies of sleek hotel chains, and Schoenberg-style scores pulsate behind most Hollywood thrillers.) Cutting-edge information-spaces will perplex their first occupants, but the most compelling designs will eventually grow more familiar, more intuitive. Users will learn over time to inhabit each new space, as though they were developing sea legs. After a few acclimations, the initial sense of disorientation will seem less intimidating, more like a challenge than an impediment. You can see this aptitude already in the generation of kids raised on video games. There's a certain fearlessness they exhibit upon entering into a new information-space. Instead of reading the manual, they'll learn the parameters in a more improvisational, hands-on fashion. (Sherry Turkle's book *Life on the Screen* has some wonderful studies of this activity.) These kids learn by doing, by experimenting, and that adventurousness comes from having cracked the code of other digital spaces in the past.

But this idea of multiple interfaces—each with its own logic, its own bylaws—also goes against the grain of interface design as we know it. Up to now, consistency has been a governing principle of the modern graphic interface. Apple gets a great deal of credit for translating the Xerox PARC desktop metaphor into a working product, but it probably deserves just as much praise for the sheer consistency of its information-spaces. For it is a basic rule of all interface design that predictability matters as much as clarity. You can have the most powerful visual metaphor in the world, but if it doesn't look the same from application to application, if the user must relearn the interface's language with each new project, then the power of that original metaphor is greatly compromised. Apple alienated some developers with its insistence that the "File" and "Edit" menus remain consistent in all applications, but that doctrinaire stand had an enormous payoff. For longtime Mac users, reaching for the "save" command is as natural, as unthinking as dialing a telephone, and the same familiarity extends to copying a block of text or printing a document. We take these conventions for granted now, but they were hard-won. It took a rigid set of interface protocols to make them possible.

This predictability—the benign sameness of shared conventions—disappears once a vibrant subculture of interface designers comes into its own. Difference and novelty are prime movers in most digital-age concerns, but in the world of interface design they can be a genuine handicap. Information architects with an eye on mainstream success will be torn between two competing drives: the siren songs of intelligibility and innovation, the desire to conform to existing conventions battling it out with the desire to push the envelope. In this one respect, traditional programmers have it easy. New features are always welcome in software programs—even if they come at the cost of memory requirements or application speed. But new interface conventions sometimes face near-insurmountable odds in their bid for acceptance, for the very reason that they happen to be new. The field of interface design, in its present incarnation, naturally inclines toward repeated patterns, the deep allure of standards, conventions, predictability. If there is a gravitational force operating within this field—the one law that cannot be resisted—it is the force of habit. If the user has learned how to do something one way, then all subsequent iterations of the software must abide by those same conventions. Never make the user learn how to do the same thing twice.

That's a good rule of thumb if ease of use is your primary goal, but if you're reaching out for more challenging, expressive possibilities, then you're certainly going to want more variety in your design options. The conflict between these two impulses—"the force of habit" versus "the shock of the new"—has played itself out in a number of typically obscure, inside-baseball debates about more adventurous interface designs. For several years now, a company called MetaTools (founded by Photoshop guru Kai Krause) has been selling high-end graphics software that sports genuinely astonishing interface design: ordinary sliders are replaced by a succession of floating orbs, each covered with a shimmering, psychedelic surface; toolbars cycle through kaleidoscope displays of random textures; scroll bars and background colors give way to fractal landscapes and Mandelbrot sets. Krause's design sensibility has its partisans and its critics: if you're not a fan of vintage Grateful Dead posters or the recursive imagery of chaos math, then you're sure to be repelled by the MetaTools interface. But here, of course, it's not just the sensibility that's at stake. Krause could dress up his windows in the visual language of Vermeer or Le Corbusier and he'd still offend some of his audience, for the simple reason that he dresses up his windows at all.

The interface medium is still young enough for those criticisms to have real merit, particularly when they're focused on basic design elements like scroll bars and close boxes. But both Apple and Microsoft have promised plug-in interface modules in their upcoming operating systems, allowing users to alter significantly the look-and-feel of their computers with third-party products. When you add to this the anything-goes design philosophy of most Web sites, it seems clear that the next decade of interface design is bound to be more diverse—and for that reason less predictable—than the preceding one. My hunch is that we should probably embrace this shift, given the aesthetic liberation it promises. A consistent look-and-feel may turn out to be one of those initial stages in the technology's development, a kind of crash course in navigating through information-space. As we slowly acclimate ourselves to the environment, too much regularity in the design may come to seem more oppressive than comforting, like a Hollywood thriller that leans too heavily on stock devices. Sure, the audience is bound to *understand* the film—it's just not clear if they'll want to sit through it. In these early days of the interface medium, consistency still reassures us. A decade from now that same consistency may feel like a shortcoming.

Metaphor Versus Simulation

One easy way to build a consistent user interface is to follow the codes and conventions of the real world. This, of course, was the fundamental logic behind Xerox PARC's desktop metaphor: if we think of the screen as a kind of mirror, reflecting the physical objects that surround us (trash cans, folders, windows), then we're already ahead of the game before we even reach for the mouse, since we can draw upon our preexisting expectations about how these objects work. In other words, the whole idea of a visual metaphor is really an extension of the more general principle of interface consistency, only this time projected out beyond the boundaries of the screen itself. The trash can works because it functions like a real-world trash can, just as a folder dutifully stores documents like a real-world folder. And yet, as we saw in chapter 2, this deference to real-world conventions has its limits. More limber, loose-fitting metaphors seem to work better than meticulous simulations, if only because the world of atoms is subject to so many restrictions that have no purchase over the world of bits.

There is also the question of the literalness of the metaphor itself. Borrowing imagery from the real world can be enormously enlightening, but if the metaphors reside too close to home, the whole onscreen experience can seem deadening, listless—like the virtual living rooms of Microsoft Bob. The fact that most computer users happen to work in corporate-style offices shouldn't be license to fashion our interfaces after those same generic spaces—if anything, the modern interface should offer an escape route from that drudgery. We don't need virtual watercoolers; we need virtual worlds where watercoolers are meaningless, worlds that serve as an antidote to the numbed repetitiveness of most information-age labor. As the representational powers of the modern interface grow, designers will be tempted to simulate the flesh-and-blood realities of office life, but the temptation should probably be resisted.

The interface design for Corbis's *Leonardo da Vinci* CD-ROM illustrates this point perfectly. One of the most elegant and informative multimedia product ever made, *Leonardo* brought together a prodigious amount of information about the Renaissance master and his epoch and, in doing so, confronted two forbidding design problems: how to represent the work of Leonardo, and how to represent the overall shape of the CD-ROM itself, with its assorted exhibits, lectures, and time lines. For the first question, the answer was simple: build a virtual museum that the user can explore, with rooms predictably divided up by genre: sketches, paintings, blueprints, and so on. It was a classic case of interface simulation: you've got an artist's work to represent onscreen, so you might as well deposit it in a sterile, austere art gallery, eight distinct rooms spun around a circular courtyard. Corbis did a wonderful job of realizing its fictional gallery-space onscreen, but the simulation seemed a little forced. Why cordon off the various strains of Leonardo's work into different rooms when the great promise of interactive media lies in the ability to make connections, to link from thought to thought and from image to image? You need separate rooms in a real-world museum, but in cyberspace they're an anomaly, a vestige held over from the world of atoms.

For the seemingly more vexing question of how to represent the entire CD-ROM, the Corbis designers opted for simplicity. Instead of erecting a fully rendered simulation, they drew upon a more poetic, if somewhat hackneyed, analogy—a tree. *Leonardo* opens out onto an oil painting of a massive oak, with two main branches trailing off from a single trunk. Waving the mouse over the image reveals a shimmering outline of the CD's contents, with the text sharpest at the cursor's tip and then fading out in all directions. The effect suggests a kind of radiance, more like shining a flashlight into a darkened room than riffling through a file cabinet. It centers the visual field masterfully, without dictating in advance where that center should be. The shape of the tree has a semantic value as well: the introductory "tours" of Leonardo and his epoch overlie the trunk, with the two branches representing the two main movements within the CD-ROM itself—the general collection of Leonardo's work, and the more detailed exhibition of the Leicester Codex (featuring the magic lens discussed in chapter 3). Presented in all its verdant splendor in that initial vista, the tree then shrinks down to icon size as you explore the rest of the disk, neatly positioned in the upper-right-hand corner of each screen. Once again, waving the cursor over the image reveals the main compartments of the overall information-space, and a single click takes you to each of them.

There is an interesting lesson here for anyone interested in the tension between metaphor and simulation in contemporary interface design. The *Leonardo* CD stores a prodigious amount of data on its laser-etched surface: four minidocumentaries on the man's life and culture, ten slide show–style exhibits on Leonardo's scientific pursuits, the massive codex display, and the art gallery. But because that particular body of information maps so nicely onto the visual metaphor of the tree, and because the icon itself appears so consistently throughout the site, it's almost impossible to lose your bearings within the disk's information-space. Your mind naturally grounds itself in one region or another, and the connections—both physical and semantic—that exist between these regions are always clear. Ironically, the most difficult space to navigate turns out to be the art gallery, where the octagonal design, with a central perspective that pivots 360 degrees, makes it difficult to sense immediately which direction you're facing (and consequently which part of the museum you're about to visit). In other words, a detailed simulation of a physical locale does a worse job of representing less information than a visual metaphor based on an abstract associative link. An art gallery might seem like an ideal conceit for a multimedia tribute to Leonardo da Vinci, but a simple tree metaphor turns out to be much more effective. In interface design, as in modern art and pulp fiction, being true to life can sometimes be a liability.

Fragmentation Versus Synthesis

One of the first Web sites I ever visited featured a giant roulette wheel on its only page; you clicked the wheel and it spun you out to a random link on the Web, sometimes buried levels deep in a site's architecture. (It usually took a few minutes of clicking around just to figure out where you were.) The Wheel struck me at the time as the perfect emblem for the Web's earliest incarnation: a page that offered nothing to its visitors but the privilege of being completely disoriented. There was no goal in this little game, no ultimate destination. You took your chances at the roulette wheel not because you were homing in on a target but because you wanted to get a little lost. Getting a little lost *was* the goal. Or at least it was more fun than knowing where you were going.

That digital roulette wheel has some formidable allies in the glittering casino of high-tech culture, though not all of them are playing at the same table. Some of them are boosters, some are neo-Luddites. Some of them claim to be innocent bystanders and nonpartisans, dragged onto the floor by a "friend with a problem." What they all share, though, is a belief—not always acknowledged, but present nonetheless—that the digital age is by definition an age of fragmentation.

This is how the story usually goes: we've begun to think in bits and packets, scattering our ideas out laterally through the infosphere, hoping for chance encounters and lucky streaks, improvising ourselves into existence along the way. Sherry Turkle embraces the "multiple selves" shuttling through online communities and MUDs, while Sven Birkerts pines for the good old days of the novel's central intelligence. David Shenk bemoans the intrusions of "data smog" into our daily lives, all those e-mail spams and news bytes diverting our attention from the real issues. Camille Paglia churns out ode after ode to her multitasking skills, typing furiously at the word processor with *Exile on Main Street* on the headphones and *Hard Copy* on the tube. Even the reigning print-design philosophies reflect this schizophrenic condition: the murky, layered look of *Raygun,* or *Wired*'s sensory-overload, "Mind Grenade" introductory pages. Beneath all the browbeating and messianism, there is this one guiding principle: zeros and ones lead inexorably toward a more fragmented experience of the world, or at least the world that comes to us over the modem and the cathode-ray tube.

It's hard not to be sympathetic to this general consensus. No one doubts that our daily lives are saturated with more data streams than at any previous point in history, and all the evidence suggests that the tide is rising. The news does come in shorter and shorter blocks (though perhaps not so short as no news at all), and the ideal spectator of most visual entertainment undoubtedly suffers from a chronic case of attention deficit disorder. The sheer number of bits that the average office worker encounters in a day is positively unfathomable. And the lush anonymity of most online encounters certainly encourages "experimenting" with your digital persona, even if most of it comes in the form of adolescent chat-room intrigue.

And yet against all that dislocation and overload and multiplicity, there is the interface. Most of the time we talk about the graphic interface as though it were a logical culmination of the digital revolution, its crowning glory, but the truth is, the interface serves largely as a *corrective* to the forces unleashed by the information age. Whenever I find myself being swayed by the fragmentation jeremiads, I like to sit down at my computer and go through the usual routines—check my e-mail, rearrange my desktop, log on to the Web—and concentrate all the while on what is *really* happening as I do these things. Because what is really happening, not on the screen but down in the innards of the machine itself, or out on the great expanses of the Internet, what is happening in that world is literally unimaginable. What is happening is that billions of tiny pulses of electricity are hurtling through silicon conduits, like an entire planet's worth of digital automobiles making their way across the grid of a single microchip. And all those pulses self-organize into larger shapes and patterns, into assembly codes, machine languages, instruction sets. Some of these ethereal languages then transform themselves into flashes of light, or audio waveforms, and depart en masse from my machine into the sprawling backbone of the Net, where they disperse into countless separate units, and then thread their way through thousands of other microchips, before reuniting at their destination.

But what happens on the screen is this: a window pops open, a dialog box appears, a bright, cheerful voice tells me that I have mail.

No news here, of course, but something profound nonetheless. The great surge of information that has swept across our society in recent years looks genuinely innocuous next to the meticulous anarchy of *real* bit-space, that netherworld that lurks in our microchips and our fiber-optic lines. But we see almost nothing of that universe because we have built such sturdy mediators to keep it separate from us, translators that make sense of what would otherwise be a blizzard of senselessness. It is undeniable that the world has never seen so many zeros and ones, so many bits and bytes of information—but by the same token, it has never been so easy to ignore them altogether, to deal only with their enormously condensed representatives on the screen. Which is why we should think of the interface, finally, as a *synthetic* form, in both senses of the word. It is a forgery of sorts, a fake landscape that passes for the real thing, and—perhaps most important—it is a form that works in the interest of synthesis, bringing disparate elements together into a cohesive whole.

Seen in this light, all that ranting about the fragmented consciousness of the digital age sounds a great deal less convincing. After all, critics have bemoaned—or championed—the accelerated pace of the present, its dislocations and divided selves, ever since the industrial age powered up in the early nineteenth century. Think of Baudelaire losing himself in the shimmering, half-lit streets of Paris, becoming a "kaleidoscope gifted with consciousness." Think of Joyce's characters bouncing back and forth between biblical references and advertising jingles. Think of Marinetti's poetry, renouncing "the 'I' in all literature" for the speed of the race car and the destructiveness of the machine gun. Conceptual turbulence—the sense of the world accelerating around you, pulling you in a thousand directions at once—is a deeply Modern tradition, with roots that go back hundreds of years. What differentiates our own historical moment is that a symbolic form has arisen designed precisely to counteract that tendency, to battle fragmentation and overload with synthesis and sense-making. The interface is a way of seeing the whole. Or, at the very least, a way of seeing its shadow, illuminated by the bright phosphor of the screen.

When I think about the gap between raw information and its numinous life on the screen—something I try to avoid doing, because it is a dark and difficult thought, more than a little like contemplating the age of the universe—the whole sensation has a strangely religious feel to it, that sense of the mind trying to reach around a vibrant (and convenient) metaphor to the wider truth that lies beyond. Cathedrals, remember, were "infinity imagined," the heavens brought down to earthly scale. The medieval mind couldn't take in the full infinity of godliness, but it could subjugate itself before the majestic spires of Chartres or Saint-Sulpice. The interface offers a comparable sidelong view onto the infosphere, half unveiling and half disappearing act. It makes information sensible to you by keeping most of it from view—for the simple reason that "most of it" is far too multitudinous to imagine in a single thought.

The spiritual resonance of interface design is not as unusual an idea as it might sound at first. Umberto Eco's compare-and-contrast exercise between operating systems and world religions circulated widely among the digital citizenry when it first appeared in 1994. Less playful critics have talked about the "technological sublime"—the Wordsworth-style reveries that come from confronting the epic expanses of information-space, the InterNIC backbone doing for a new generation of aesthetes what the Matterhorn did for the Romantics nearly two centuries before. As I write, the Silicon Valley start-ups are devising new types of onscreen "avatars"—digital creatures that represent you in your virtual habitats—borrowing the Buddhist term for angels. For me, the most moving rendition of this theme comes almost as an aside in Thomas Pynchon's 1990 novel, *Vineland:*

> If patterns of ones and zeroes were "like" patterns of human lives and deaths, if everything

about an individual could be represented in a computer record by a long string of ones and zeroes, then what kind of creature would be represented by a long string of lives and deaths? It would have to be up one level at least—an angel, a minor god, something in a UFO. It would take eight human lives and deaths just to form one character in this being's name—its complete dossier might take up a considerable piece of the history of the world . . .

None of this is to suggest that there are genuine religious values to be found on the Net or in our microprocessors. While certain strains of New Age mysticism seem to have embraced digital technology, for the most part the modern computer is a deeply secular invention. Still, the act of comprehending an infinite universe of data through the figureheads and symbolic gestures of the interface, the whole project of "infinity imagined"—this experience runs parallel to the metaphors and sense-making narratives of most organized religions. They share a similar "structure of feeling," in Raymond Williams's term, the sense of a disordered universe made orderly again by the power of metaphor. And in a world that increasingly lays its tributes at the great altar of information, where the "symbolic analysts" and digital visionaries sometimes seem like a new caste of priests and prophets, then perhaps the visual metaphors of interface design will eventually acquire a richness and profundity that rival those of Hinduism or Christianity, without crossing over into genuine theology. The empire of Byzantium ruled much of southern Europe and eastern Asia for nearly a thousand years, but during the eighth and ninth centuries, the regime was locked in a vicious internecine war over the role of icons in orthodox worship. (The modern word *iconoclasm* derives from this debate.) Was the icon a suitable stand-in for the sacred—or was it a perversion, a false idol? Did it bring us closer to the heavens, or condemn us to hell? You can hear the same melody today in the great symphony of high-tech culture—fluttering softly in the background, of course, and transposed into a secular key, but it is the same melody nonetheless. Whatever else may befall the digital world in the coming years, that spiritual refrain is bound to grow louder.

I wrote in the Preface that I saw this as a "secular" book, a middle ground between the dual religions of techno-boosterism and the Luddite reaction, and for the most part, I have tried to stay true to that original vision. If there is a spiritual dimension to the interface medium, it has nothing to do with dogma or unapologetic mysticism. It has nothing to do with believing—or not believing—in God. It has to do more with the general structure of trying to think about something that is too big to think, and the devices we build for ourselves to help us complete the thought. Other forms in history have taken on similar quandaries: Dickens and Balzac condensed down the teeming masses of the modern metropolis into five hundred pages; a radio station here in New York City regularly announces: "Give us twenty-two minutes and we'll give you the world." But these forms at least have the luxury of representing a world that can be experienced through other means. You could stroll along the Seine or take a gander at Chancery to experience the worlds of Balzac and Dickens more viscerally. The novel made sense of social movements that transcended the scale of individual lives—industrialization, urban population explosions, epidemics—but you could still venture out into the city streets to encounter those trends in their day-by-day manifestations.

With a few momentous exceptions, the sense-making apparatus of religious belief has not had it so easy. Experiencing godhead usually involves some kind of mediation, if only because most humans accept the idea that a direct encounter might blow a few fuses in the act. (There's a limited-load-capacity clause written into most sacred texts for good reason.) This is where the modern interface resonates so powerfully with the customs and pageantry of organized faith. Both are imaginative systems predicated on a world ruled by invisible forces, forces made sensuous only through the luminous icons and rituals of faith. Interface designers talk about the "user illusion," but there is also a strong measure of "suspended disbelief" in the modern desktop—which, if you cancel out the negatives, leaves you with old-fashioned belief. This is probably how it should be.

The interface came into the world under the cloak of efficiency, and it is now emerging—chrysalis-style—as a genuine art form. All this in less than half a century of innovation. Who can tell what awaits us in the next fifty years? The religious analogy seems less rhetorical when measured against that scale. Even today, there's an undeniably enchanted quality to the icons on our screens, like a crucifix or the lives of the saints. We can't predict how far that enchantment will extend itself in the next century, but its potential scope should not be underestimated. Our interfaces are stories we tell ourselves to ward off the senselessness, memory palaces built out of silicon and light. They will continue to change the way we imagine information, and in doing so they are bound to change us as well—for the better *and* for the worse. How could it be otherwise?

interface culture

NOTES

Preface: Electric Speed

PAGE 5: *"understanding of causes"*

McLuhan, Marshall, *Understanding Media: The Extensions of Man* (Cambridge: MIT Press, 1996), 353.

Chapter 1—Bitmapping: An Introduction

PAGE 11–12: *The Greek poet Simonides*

Woolley, Benjamin, *Virtual Worlds* (London: Penguin Books), 138.

PAGE 13: *"books in the sixteenth century"*

Spence, Jonathan D, *The Memory Palace of Matteo Ricci* (New York: Penguin, 1984), 13.

PAGE 14: *"literary machines"*

Nelson, Ted, *Literary Machines* (Self-published, 1981).

PAGE 16: *"head like an elephant"*

Dickens, Charles, *Hard Times,* (New York: Penguin, 1985), 65.

PAGE 18: *"cognitively map"*

Lynch, Kevin, *The Image and the City,* (Cambridge: MIT Press, 1960).

PAGE 23: "mind processes information"

Rheingold, Howard, *Tools for Thought* (New York: Simon and Schuster, 1985).

PAGE 23: *"quick continuous jolts"*

Celine, Louis-Fernand, *Journey to the End of Night,* trans. Ralph Manheim (New York: New Directions, 1983), 194.

PAGE 24: *"metal of motors"*

Marinetti, Filippo Tommaso, *The Futurist Cookbook* (New York: Chronicle Books), 89.

PAGE 24: *"city of bits"*

Mitchell, William J., *City of Bits* (Cambridge: MIT Press, 1995).

PAGE 30: *"its own needs and agendas"*

Rushkoff, Douglas, *Media Virus* (New York: Ballantine, 1996), 23.

PAGE 31: *"endurance and labor"*

Eliot, George, *Middlemarch* (New York: Penguin, 1965), 846.

PAGE 39: *stream-of-consciousness*

For a more in-depth analysis of the evolution of Joyce's device, see Franco Moretti's essay, "On Literary Evolution," from his collection *Signs Taken for Wonders.*

PAGE 40: *"different traditions"*

Gould, Stephen Jay, *Full House* (New York: Harmony Books, 1996), 221.

Chapter 2—The Desktop

PAGE 47: *"top of the pile"*

Levy, Steven, *Insanely Great* (New York: Penguin, 1994), 61.

PAGE 50: "an entire generation."

Kay, Alan, "User Interface: A Personal View." from Laurel, Brenda. ed, *The Art of Human-Computer Interface Design* (New York: Addison-Wesley, 1990), 189.

PAGE 54: *"think and type"*

Pennington, Harvard, "Of Mice, Windows, Icons, and Men," *Creative Computing.* Vol. 10, No. 11, pg. 215.

PAGE 55: *"designing them as well"*

Forbes, February 13, 1984.

PAGE 55: *"true desktop environment"*

Bonner, Paul. "The Desktop Environment." *Personal Computing,* August, 1984, 72.

PAGE 59: *"Clearly not."*

Kay, pg. 199.

PAGE 64: *Robert Moses*

For more on the connection between Hassmann and Moses, see Marshall Berman's *All That Is Solid Melts Into Air.*

PAGE 67: *"kaleidoscope of consciousness"*

Baudelaire, Charles, *The Painter of Modern Life and Other Essays,* trans. Jonathan Mayne (New York: Phaidon Press, 1964), 8.

Chapter 3—Windows

PAGE 83: *"our many selves"*

Turkle, Sherry, *Life on the Screen* (New York: Simon and Schuster, 1996), 168.

PAGE 87: *White Mythologies*

Derrida, Jacques, *Margins of Philosophy*, trans. Alan Bass (Chicago: University of Chicago Press, 1984).

PAGE 90: *"our memory does"*

Borges, Jorge Luis, *A Personal Anthology* (New York: Grove Press, 1967), 51.

PAGE 91: *"wholly different purpose"*

Darwin, Charles, *The Origin of Species* (New York: Random House, 1993), 210.

Chapter 4—Links

PAGE 112: *"what was it?"*

Dickens, Charles, *Great Expectations* (New York: Washington

Square Press, 1964), 228.

PAGE 114: *"Estella's Mother"*

Dickens, *Great Expectations,* 374.

PAGE 117: "square-rigged ships"

All quotes from "As We May Think" are from the digital version of the essay, stored at The Atlantic Monthly Web site, at http://www.theatlantic.com. The original essay ran in the July 1945 issue of the print magazine.

PAGE 125: *"reading it ends"*

Joyce, Michael, *Afternoon, A Story* (Watertown: Eastgate Systems, 1993).

PAGE 131: *"In the new infomockracy"*

All quotations from Suck are from their web site, accessible at http://www.suck.com

PAGE 136: *"Or just after."*

Stevens, Wallace, *Selected Poems* (New York: Vintage, 1967), 20.

Chapter 5—Text

PAGE 147: *"never imagined"*

Gibson, William, *Neuromancer* (New York: Ace Books, 1984), 23.

PAGE 151: *"topsight"*

Gelenter, David, *Mirror Worlds* (New York: Oxford, 1991).

PAGE 158: *"onstage together"*

Dolnick, Edward, "The Ghost's Vocabulary," *The Atlantic Monthly,* October 1991.

Chapter 6—Agents

PAGE 178: *"documentary novel."*

McLuhan, pg. 121.

PAGE 181: *"Brain Tennis"*

Quotations from "Agents of Alienation" are from the digital version of the document, accessible at http://www.voyagerco.com. Excerpts from the BrainTennis discussion are available at http://www.braintennis.com.

PAGE 190: *"rain gear"*

Editors, "Push!" *Wired Magazine,* February 1997.

Conclusion: Infinity Imagined

PAGE 208: *upgrade to the telephone*

Baldwin, Neil, *Edison: Inventing the Century* (New York: Hyperion, 1995), 85.

PAGE 217: *"the old experience"*

Eliot, George, *The Mill on the Floss* (New York: Penguin, 1986), 230.

PAGE 219: *"vanish altogether"*

Birkerts, Sven, *The Gutenberg Elegies* (Boston: Faber and Faber, 1994), 202.

PAGE 240: *"history of the world"*

Pynchon, Thomas, *Vineland* (New York: Viking, 1990), 45.

From the beginning I have thought of *Interface Culture* as a book of links, and nowhere is this connectedness more apparent to me than in the wonderful confluence of writers, colleagues, family, and friends that helped make it possible. I have tried to do some justice to the book's forebears in the text itself, and in the appendix. But just to set the record straight, Interface Culture was composed under the influence of the following writers and designers, to whom I owe an enormous debt of gratitude: Daniel Boorstin, Marshall McLuhan, Raymond Williams, Alvin Toffler, Donna Harraway, Kevin Kelly, Howard Rheingold, Benjamin Woolley, Sven Birkerts, Umberto Eco, Steven Levy, Brenda Laurel, Bruce Tognazzini, Joy Mountford, Mark Pesce, Joe Belfiore, and Edward Tufte. I am particularly indebted to three professors who tolerated me through my undergraduate and graduate years. Neil Lazarus and Robert Scholes helped me understand the possibilities of cultural criticism. Back in my grad school days, Franco Moretti used to call me now and again for help with his word processor, and so he may be surprised to find that his work has had such a profound influence on a book about technology.

The list of friends and colleagues who contributed to my thinking over the past years is endless, but here's a start: David Bennahum, Nicholas Butterworth, Matt Buckley, Denise Caruso, Rory Kennedy, Annie Keating, Liz Garbus, Rufus Griscom, Neil Levi, David Lipscomb, David Marglin, Ceridwen Morris, Jack Murnighan, Alex Ross, Alex Star, Elizabeth Schmidt, Mark Tribe, Alexandra Valenti, Gary Wolf, and my extraordinary colleagues at work—Amanda Griscom, Austin Bunn, and Irwin Chen. Two other members of the good ship FEED deserve special mention here. For the past two years, Stefanie Syman has been my editor, sounding board, and side-walk media mogul; all working friendships should be as collab-orative and sustaining as ours. The publication of this book means that I owe Sam Lipsyte a dinner, but as a writer, editor, and friend my debt to him extends well beyond that.

Other conspirators who deserve special indict-ment: Eric Liftin taught me more about the intersection of interface and architecture than anyone; JJ Gifford lent me a chapter title and nearly a decade of high-tech wisdom; Jay Haynes and I spent countless late nights rolling over the Mac vs. Windows debate and other interface imbroglios. (Jim Haynes joined in for a few of those, as I recall.) For reasons too complicated to explain here, this book probably wouldn't have happened without the friendship and hospitality of Ruthie and Roo Rogers. Penny Lewis—my research assistant and dear friend—provided a sense of history, both on the job and off. My agent, Lydia Wills, somehow saw the merits of this project when it was a thinly disguised doctoral dissertation on Charles Dickens. All editors should be as perceptive and supportive as

Eamon Dolan; some of my favorite sections of this book came directly from his suggestions.

Finally, Alexa Robinson sustained me through four grueling months of work with sweetness, compassion, and a critical eye. No one shaped my thinking more profoundly—from the book's half-baked origins three years ago to its current, fully-baked version. May there be many more collaborations in the years to come. As for my family—my parents, grandparents, sisters, and brother—they know all too well how much I have leaned on them for advice, emotional support, and intellectual stimulation, but they may not be aware of how much gratitude I feel for that sustenance. Truly happy families may not be all alike, but they are rarities in this world, and anyone lucky enough to be born into one should consider himself blessed. Think of this book as one small attempt to count those blessings.

•

a
c
k
n
o
w
l
e
d
g
m
e
n
t
s

INDEX

251

·

i
n
d
e
x